FROM STARS TO STALAGMITES

How Everything Connects

FROM **STARS** TO **STALAGMITES**

How Everything Connects

Paul S Braterman

University of Glasgow, UK

World Scientific

NEW JERSEY · LONDON · SINGAPORE · BEIJING · SHANGHAI · HONG KONG · TAIPEI · CHENNAI

Published by

World Scientific Publishing Co. Pte. Ltd.

5 Toh Tuck Link, Singapore 596224

USA office: 27 Warren Street, Suite 401-402, Hackensack, NJ 07601

UK office: 57 Shelton Street, Covent Garden, London WC2H 9HE

British Library Cataloguing-in-Publication Data
A catalogue record for this book is available from the British Library.

FROM STARS TO STALAGMITES
How Everything Connects

34.20

540
Bra

ISBN-13 978-981-4324-97-7 (pbk)
ISBN-10 981-4324-97-3 (pbk)

Typeset by Stallion Press
Email: enquiries@stallionpress.com

Printed in Singapore by B & Jo Enterprise Pte Ltd

Contents

Acknowledgements

It is a pleasure to thank the many people who have helped make this book possible: my darling wife Rebecca, without whom none of this would have been accomplished; my students and colleagues over many years, and in particular my students from TAMS, the Texas Academy of Mathematics and Science in the University of North Texas, for whom much of this material was originally written; Roald Hoffmann, Peter Atkins, and Steve Newton for early encouragement; my colleagues at CESAME, the [New Mexico] Coalition for Excellence in Science and Mathematics Education, and BCSE, the British Centre for Science Education; numerous individuals for helpful comments, reading parts of the text, or supplying materials, including Ariel Anbar, Mark Boslough, Laura Braterman, Professor D. Chakrabarti, John Crooks, Tais Dahl, Alwyn Davies, John Fleck, Haim Harari, M. Kim Johnson, Jim Kasting, Bernard Knapp, Nick Lane, Jim Marshall, Diana Mason, Steve Mojzsis, Ian Orland, Philipp Podsiadlowski, Thilo Rehren, Gary Rendsburg, Bryan Reuben, Merav Segal, Fritz Stern, John Wiltshire, and Jaime Wisniak; the ever-helpful and well-informed library staff at UNT and Glasgow; and V.K. Sanjeed and the editorial team at World Scientific for transforming my raw manuscript into the volume you are now reading.

Introduction

If we assume we've arrived, we stop searching, and we stop developing.

Dame Jocelyn Bell Burnell,
discoverer of the first known pulsar

Science is a journey. What we know cannot be divorced from how we came to know it. We make progress, we overcome obstacles, we pursue blind alleys, take lengthy detours, make breakthroughs, and are rewarded with fresh vistas.

When discussing the tangled paths that have led to our present vantage point, I have tried to strike a balance between the wisdom of hindsight, and discussing complex issues as they appeared to their discoverers. I have presented topics in an order chosen for ease of understanding, which is rarely the same as the order in which the discoveries were made. At the same time, I have tried not to pick heroes and villains, or even winners and losers, on the basis of what we know today. Science is disputatious, errors can be fruitful, and the opponents of a new concept help refine it through their very criticisms.

This book is unique in its range of subject matter, its level of treatment, and the manner in which the material is related to its historical and intellectual context. The topics chosen are of fundamental importance, of topical interest, or illustrate particularly important interactions between scientific developments and society. Each topic is presented within its historical context, and with descriptions of the leading figures and what was at stake for them. Finally, the underlying science is discussed in non-technical language in a context that makes it meaningful, and this principle is applied to such "difficult" topics as thermodynamics and the nature of chemical bonding.

The book is all about how things connect, in three rather different senses. Richard Feynman once selected, as the single most important statement in science, that everything is made of atoms. It follows that the properties of everything depend on how those atoms are connected to each other. At the same time, our knowledge of this topic, as of every other, is the hard-won product of human endeavour. Science is not a mass of dead data, but a vibrant cultural activity. There are reasons why science has flourished in some periods but not others, why scientists chose to work on the particular problems that they did, and why one solution, and not necessarily the best solution, seemed to them to be preferable to another. There are also obvious connections between scientific discoveries and historical outcomes, although these outcomes may be very different from what the discoverers had in mind. So that is the second kind of connection, the connection between science and society. The third kind of connection is within science itself. Ideas generated in one context prove fertile in another, and reality does not respect the arbitrary divisions we make between different "subjects".

The chapters cover a range of topics that seem to me particularly important or engaging. The book starts with the age of the earth, and concludes with the life cycle of stars. In between, we have atoms (and molecules) old and new, the discovery of the noble gases, synthetic fertilisers and explosives, the ozone hole mystery and how it was solved, reading the climate record, the extraction of metals, and how the greenhouse effect on climate really works. The different ways in which atoms and molecules interact with each other are described in context as they arise. A chapter in praise of uncertainty leads on to the "fuzziness" and sharing of electrons, and from there to molecular shape, grass-green and blood-red, the wetness of water, and molecular recognition as the basis of life.

Within each chapter, one particular development or issue is discussed in its historical and social context, with enough background to explain the beliefs and actions of the participants, what was or is at stake, and why it matters. Each chapter can be read on its own, but there are repeated connections between them, and I make a point of drawing the reader's attention to these. That is because there are some general unifying concepts, such as atoms, energy, entropy, and shape, that make repeated appearances, and with these I have sought to use the novelist's devices of interlocking narratives and emerging themes. Chapters end with a summary of the main points covered, and the book as a whole contains a glossary of technical or unusual terms. The book concludes with endnotes that give suggestions for further reading, detailed references where it seemed to me important to back up particular statements, chemical equations for the processes discussed in the text, and a deeper discussion of some of the more complex issues. I have used everyday language, and made the technical content as simple as possible, but — as Einstein supposedly warned — no simpler.

1 The Age of the Earth — An Age-Old Question

The age or agelessness of the Earth has been discussed in philosophy and religion from antiquity. Modern scientific estimates emerged through the nineteenth century with the development of geology, leading to our present estimates based on radioactive dating, developed in the early 20th century. These great ages seemed to be in conflict with the principles of thermodynamics. It was thought that the Sun's energy could only have been enough to shine for some 20 million years, but Rutherford estimated the age of Cambrian rocks at over 500 million years.

Such an age can be reconciled with the laws of physics only if we take into account the effects of radioactive decay generating heat in the Earth's crust, and of the fusion of hydrogen to helium in the Sun.

Our best current estimates are somewhat over 4 billion years for the oldest surviving rock grains on Earth, and 4.57 billion years for the Solar System. The geological cycles of erosion and uplift are driven by plate tectonics, which in turn is a result of the outward flow of heat generated from radioactivity. The Sun itself will run out of hydrogen some 4 billion years from now.

Pre-Scientific Theories; in the Beginning …

How old do we now think the Earth is? How did we get to that position? How certain are our conclusions? What can we learn about methods of scientific and other thought from this example?

The first question is, does the Earth really have an age, or has it always existed? The traditional Hindu and Buddhist belief is in an eternal, cyclical time. The Maya of Central America may have had the same idea.[1] There is nothing unscientific about this view, although we now know it to be wrong. It assumes (correctly) that the laws of nature do not change and (incorrectly, but in the absence of evidence to the contrary) that the world in which they operate does not change very much either. This belief in an ageless Earth was shared by Pythagoras (of the right-angled triangle theorem) and his followers, and passed on through them to Plato and Aristotle.

The opposite view, that the Earth was created at some specific time, is found in Sumerian legends, and was passed on with modifications to the Hebrews and also to the Greeks. In this view, the origin of the Earth was an event in history, or rather an event defining the beginning of history, when Earth was created by the Gods (or, for the Hebrews, by God).

The pre-Socratic philosopher Democritus[2] considered that basically there were only two kinds of thing in existence, atoms and the void. Both of these were eternal, indestructible, and infinite in extent and number; thus for him there was no such thing as the creation of the Universe as a whole. The atoms have rearranged themselves over time within the void, perhaps quite dramatically, but both atoms and void have existed for ever. However, the creation of our Earth is quite another matter. The Roman poet Lucretius, who drew his inspiration from Democritus and his followers, argued that it could not be infinitely old, for if it were, why do we have no historical records older than the time of the Trojan war (ca. 1200 BC, or not much more than a thousand years before his own time)? So at this stage we have the rationalists and the religious conservatives in the Mediterranean world agreeing on a relatively young Earth, with the mystics of both East and West, and the philosophers they had influenced, arguing for one that is infinitely older.

If one regards the Old Testament as a historical account, then it should be possible to estimate the date of creation from its chronology. In the 4th century AD the compilers of the Talmud carried out this calculation, and came up with the biblical date of creation as 3760 BC. This remains the "zero" for the Jewish ritual year. The calculation was repeated in the 17th century by Bishop Ussher, who made different estimates of the gaps in the narrative and the interpretation of some verses, and came up with the date of creation as 4004 BC. (Why 4004 rather than Martin Luther's 4000? Because according to the Gospels, Jesus must have been born while King Herod was alive, and Herod died in 4 BC.)

However, long before Bishop Ussher's time, some had begun to question such a literal-minded use of scripture. One instructive example is Maimonides (1135–1204 AD). He was the son of a rabbi, born in Moorish Spain. During its golden age, Moorish (i.e. Muslim-ruled) Spain had been characterised by great tolerance and openness, and was one of the gateways through which classical learning re-entered Christian Europe as the latter emerged from its Dark Ages. At the time Maimonides was born, this openness was being eroded by religious dogma, and when Cordoba, his birthplace, fell under the control of an intolerant Islamic sect, his family moved to Cairo. Moorish Spain, split by dynastic and sectarian disputes, continued its slow decline, coming to an end in 1492 when the Christians took Granada, the last Moorish stronghold. Thus do intolerant societies impoverish themselves by driving out the independent thinkers who form such a valuable if uncomfortable resource.

In Cairo, at that time a great centre of learning, Maimonides did very well as a doctor. He became physician to Sultan Saladin, and was offered (and refused) a similar job in the court of King Richard the Lionheart of England. He was an authority on herbs and medicine, a leading biblical commentator, and consultant and codifier of Jewish religious law. He also wrote a text on logic, and a monograph on poisons, although it is mainly as a philosopher that he is remembered today.

Maimonides' best known work now is the "Guide to the Perplexed", first written in Arabic. In this, he attempted to reconcile his own religious orthodoxy with the best scientific and philosophical thought of his time. He believed as firmly as any present day Fundamentalist that every word in the Bible was divinely inspired, but regarded the sacred text as metaphorical by its very nature, since it had to be written in language that humans could understand. Specifically, he used arguments from the text itself to show that literalism was untenable. The account of creation in particular must not be taken literally. For instance, "day" could not have its common meaning if Sun and Moon were not created until the fourth day. He also argued that laws of nature were not fixed until the end of the Creation, so that the very meaning of "time" at that stage was unclear. For what he calls the "ill-informed theologian ... possessing no knowledge of the nature of things," who takes at face value "statements which involve impossibilities", Maimonides has nothing but contempt. The issues that Maimonides grappled with here are regarded as of merely historical interest in Western Europe, but are still very much alive in some areas of the United States and, ironically, in the Muslim world.

Maimonides powerfully influenced Thomas Aquinas, and through him Western thought in general. Aquinas, in the section of his Summa Theologica devoted to what he called "the work of the Six Days", explicitly admitted that a wide range of different interpretations of Genesis was possible, and advised that "… since Holy Scripture can be explained in a multiplicity of senses, one should adhere to a particular explanation, only in such measure as to be ready to abandon it, if it be proved with certainty to be false; lest Holy Scripture be exposed to the ridicule of unbelievers, and obstacles be placed to their believing". Nothing new, it seems, about literalist zealots embarrassing their co-religionists.

The Transition to Modern Thinking

Observational arguments in favour of the world's antiquity go back at least a century before Maimonides. Al-Biruni and Ibn-Sina (Avicenna), Persian scholars active in the first half of the 11th century AD, both realised that river valleys were the result of erosion by the streams that flowed through them, and that this was a process that must have taken an enormous length of time. Since there is nothing in the Koran of direct relevance to the age of the Earth, there were no theological obstacles to such reasoning. Al-Biruni went even further, and recognised the valley of the Indus as an alluvial plain, with coarser debris incorporated further upstream, where the force of the water was greater.

For comparable insight in Christian Europe, we must wait for Leonardo da Vinci. Leonardo described fossil shells found far inland, high up in the Apennine mountains that form the backbone of Italy. He reasoned that they could not possibly have got there in the forty days of Noah's flood, and inferred (correctly) that land and sea had changed places over much longer periods of time. Prudently, he kept these views to himself, and the relevant writings were not discovered until the 19th century.

Some 200 years later, Robert Hooke, a contemporary of Newton,[3] predicted that the study of fossils would provide the basis for a new chronology, "which, in all probability, will far antedate all the most ancient monuments of the world, even the very pyramids". Newton himself suggested that Earth had rotated much more slowly during the Creation. The philosophers Hobbes (1588–1679), and Spinoza (1632–1677) saw scripture as an evolved human document. Benoît de Maillet (1656–1738) estimated that strata are laid down at a rate of one metre every thousand years, and used this to try to date fossil organisms from the depth at which their remains occur. (He also noted that humans, and different kinds of animals, have body structures with common features, and from this, over a century before Darwin, he inferred evolution from a common ancestor.)

The French naturalist Buffon applied scientific reasoning in 1749 to the age of the Earth, but retracted under pressure from theologians. In 1774 he published "Introduction to the history of minerals", in which he argued "Just as in civil history we decipher ancient inscriptions in order to fix the dates of events, so in natural history one must dig through the archives of the world, extract ancient relics, and assemble again those indications that can carry us back to the different Ages of Nature". He also stated (cf Maimonides) that Genesis had been given to the uninformed, who could not have comprehended large times, and that "day" must therefore have had an allegorical meaning. He then estimated the age of the Earth, using estimates of cooling rates, at around 75,000 years. His calculation was wildly wrong, but the abandonment of biblical literalism, and his treatment of the age of the Earth as a problem in science, are thoroughly modern.

The Beginnings of Geological Dating

From 1750 onwards, geologists came to realise that the stones of central France were the same as those of the area round Vesuvius. They could even see where flowing lava had scorched the older rock beneath it, and trace the flow back to its parent volcanic cone. The basalt plateaus and cone peaks of the Massif Central were evidence of an ancient period of volcanism of which history knew nothing.

We can date the emergence of the modern science of geology to James Hutton's writings in 1785 and 1788. Hutton lived in Edinburgh, and was part of the "Scottish Enlightenment", along with the chemist Joseph Black (who isolated carbon dioxide by heating limestone), the philosopher David Hume, James Watt of steam engine fame, the economist Adam Smith, and the physician William Prout. Prout is remembered for discovering that the digestive fluid is mainly hydrochloric acid, and for hypothesising that all atoms were made out of hydrogen, since so many of their atomic weights were very close to whole numbers.[4] Hutton himself originally trained as a lawyer and physician, but when he inherited a farm from his father he became interested in the nature of the land and what lay underneath it. This interest would have been further stimulated by his involvement in the building of the Forth and Clyde Canal, where he was shareholder, director, and (as we would now say) scientific adviser.

It is to Hutton that we owe our understanding of unconformities as evidence for long cycles of deposition, erosion, and re-deposition (Figure 1.1). The strata, or layers, show how sedimentary rocks were originally laid down. An unconformity is a place where the rocks lying on top of each other do not match. There is a sudden change in the exact nature of the rock, the strata do not fit together, and often we find that the rocks above the unconformity have a different tilt from those underneath. The

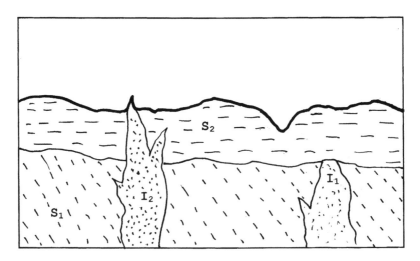

Figure 1.1. Sedimentary rocks with unconformity and igneous intrusions. S_1, lower sedimentary rock; S_2 upper sedimentary rock; I_1 older igneous rock; I_2 younger igneous rock. The ages are clearly in the order $S_1 > I_1 > S_2 > I_2$.

inference is that the first set of strata were laid down, consolidated, tilted, and eroded before the next set formed. Even more striking is the case where igneous rock (granite or basalt) has penetrated the first rock to the level of the unconformity but does not enter the second; here we can infer that the first rock was laid down before the igneous intrusion, but that the second was formed later. We now find this kind of argument particularly useful since the methods developed by Rutherford and his followers in the 20th century (see below) apply most directly to igneous rocks.

It is to Hutton also that we owe the modern scientific concept of "deep time". We can see how slowly mountains erode, or new strata are laid down in river valleys, and it takes but little thought to see that repeated cycles of this kind must have occupied a length of time far greater than all of human history. Hutton's ideas were developed further by (Sir) Charles Lyell in the 1830s. Lyell had the great advantage of living and working at the time when railways, as well as canals, were spreading over the countryside, and when coal mining was expanding. This gave him the opportunity to inspect freshly exposed rock faces, as new mining shafts were sunk, and as the new canals, and railroads whose steam-driven locomotives could not tolerate steep gradients, had to be provided with cuttings. He even had available to him the earliest geological map,[5] published in 1815 by the canal engineer William Smith, showing the existence of similar strata separated at times by hundreds of miles.

Lyell used his detailed observations to develop the modern science of geology, and attempted to use the rates of geological processes to estimate the age of the various rocks. For example, he attempted to estimate the age of Mount Vesuvius from its height, its frequency of eruption, and the average thickness of the excavated lava beds. It is to Lyell that we owe the complex set of ideas that we now call "*uniformitarianism*". There are several separate ideas bundled together under that blanket. These ideas are not all equally obvious, and some are on occasion demonstrably false.

The first idea is that the laws of nature have not changed, and apart from questions about the very earliest state of the Universe, scientists pretty much accept this. At one time this was merely an assumption, but now (see Chapter 16) we have evidence to support it, and the constancy throughout time and space of the laws of nature has developed from a philosophical assumption to an observed fact.

The second idea is that the kinds of processes at work in geology have not changed, and we now have rock strata going back to some 4.03 billion years before present, and individual rock grains (zircons) up to almost 0.2 billion years beyond that,[6] whose chemistry and texture support this belief. The third idea is that the rates of processes have not changed; this is true only if we paint with a *very* broad brush. There are periods, for instance, like the present, when mountains are rising relatively quickly. The northward movement of the plates carrying India and Africa brought them into collision with the Eurasian plate some 20 million years ago, forcing up the Himalayas, Alps, and Pyrenees, which are still rising. There have been other periods when land masses were not in collision, and mountain building was dormant. Erosion rates also vary dramatically with conditions. The steep valleys that dominate the Scottish Highlands, Yosemite, and the fiords of Alaska and Norway were gouged out by glaciers very rapidly, as these things go. Topography also makes a difference, with higher ground being more exposed to erosion. Neglect of such factors can lead to quite embarrassing mistakes, such as Darwin's estimate from present erosion rates that the valley between the North and South Downs in south-east England took 300 million years to form, whereas we now know that the chalk it cuts through is only 60 million years old. Thus the assumption of uniform rates is at best unhelpful, at worst, actively misleading.

The fourth and last idea is that on a grand scale the Earth itself, at least in its physical features, has not changed. This, as we now know, is simply not true. For example, we know from the chemical composition of Precambrian rocks that there was relatively little oxygen in the Earth's atmosphere until sometime around 2.5 billion years ago.[7] Another significant change comes from the evolution of major plant forms, which speed up the weathering of rocks, but slow the erosion of soil. However, Lyell was in no position to know these things. Most of the easily accessible rocks are from the last 500 million years or so, during which life has evolved dramatically, and continents have shifted, joined, and separated, but the physical state of the Earth's atmosphere has not changed all that much. Hutton and Lyell, whose information derived mainly from the geology of Great Britain, had no reason to be aware of the extent of earlier changes. This makes the Earth, by default, indefinitely old, with geology proceeding in cycles of erosion and uplift. To quote Hutton, "The result therefore of our present enquiry is that we find no vestige of a beginning — no prospect of an end."

The Conflict with Thermodynamics

Other developments, however, were to undermine this view. I have already mentioned steam engines and railways. Science in the mid 19th century was much occupied with matters concerning work and energy, and the efficiency of heat engines. This period saw the development of a new subject, *thermodynamics*, dedicated to such matters. One of the most fundamental results of thermodynamics (the First Law) is that energy is conserved. Another (the Second Law) is, that since energy tends to spread out and degrade irreversibly over time, there could be no such thing as a perpetual motion machine. Any real process, and certainly such a process as the uplift and erosion of the Earth, is operating against friction, with overall irreversible degradation of energy into heat, and this is something that cannot continue on its own indefinitely. Yet the Earth, as seen by Hutton and Lyell, appeared to be just one such machine, running through cycles of uplift and erosion with no visible source of energy to drive the process. Conflict between the thermodynamicists and the geologists was inevitable.

William Thomson, Lord Kelvin,[8] in whose honour the absolute temperature scale is now named, was among the most distinguished scientists of the late 19th century. His work straddled the boundary between pure and applied research. Among other things, he played a major role in establishing the relationships between heat, work, and electricity, worked out the theory for how much information (as we would now say) could be carried by the first submarine cable, and improved the form of the compass and the methods of navigation. He was appointed Professor of Natural Philosophy (i.e. Physics) at Glasgow University when he was 22, and held that Chair for more than 50 years.

Kelvin was interested in the age of the Earth, considered as a problem in physics, from a very early stage. It was the subject of a prize undergraduate essay, and also of his inaugural lecture at Glasgow, now unfortunately lost. He was also a sharp critic of the science of geology as it was developing. He argued (correctly) that extreme uniformitarianism was not compatible with the laws of physics. Things must have been very different at some time in the past, and would be different again in the future. The Earth was losing heat and must have once been a molten ball. The Sun was emitting energy, could not have been there forever, and must eventually run out of energy, plunging the Earth into utter cold and darkness.

In a lengthy series of publications,[9] Kelvin attempted to quantify these general objections. He developed a way of estimating the age of the Earth's solid crust from cooling arguments. It is hot down a mine, and the deeper you go, the hotter. If you could go down deep enough, you would, at a depth of some miles, reach the Earth's *mantle*, where the rock is actually molten. So if we have cooler rocks on top and hotter rocks lower down, heat must be flowing up through the rocks from the centre outwards. Knowing how fast the temperature increases as we go down, and how effectively the rocks of the crust conduct heat, Kelvin calculated how fast the Earth was losing heat. Where was this heat coming from? Kelvin thought he had the answer. He assumed (correctly) that the Earth was originally molten, and that heat must have dissipated as the Earth's rocks solidified from an originally molten state (the opposite kind of process to ice absorbing heat as it melts). From an estimate of the thickness of the solid rock layer (the *crust*), and from measurements of how much heat it takes to melt a given amount of rock, he was able to estimate how much heat has been given out by this process of solidification. Then, by running the model backwards through time, he calculated that the thickness of the Earth's solid crust corresponded to 100 million years. At this date before the present, all the rocks now on the surface would have been molten, and this, according to his argument, is therefore an absolute upper limit on the age of the solid crust of the Earth.

Throughout the 19th century, unaware of this approaching crisis, geologists worked away at establishing our familiar geological column.[10] They worked out the order of the strata from which lay on top of which others, and later from the complexity of the fossils they contained, and made estimates of the duration of each geological period from the thicknesses of its best preserved sediments. Not realising that they had only a very incomplete part of the total depositional record, they came up with an estimated age of around 100 million years upwards, in tolerable agreement with Kelvin.

The age of the Sun presented a much more serious problem. We know how large the Sun is, how far away, and how much solar energy reaches us. From this, it is relatively straightforward to calculate its total energy output. Where is this energy coming from? Not from any chemical process, for no chemical process is energetic enough. So Kelvin, building on suggestions by Helmholtz and others, suggested that a more useful source might be the gravitational energy released during the Sun's formation. Knowing the total mass of the Sun, and using Newton's Laws of gravitational attraction, Kelvin could work out how much energy must have been given out by this process. This would first be converted into the kinetic energy of the infalling matter, and that kinetic energy would then by well-known physical processes be converted to heat and ultimately to light, all in strict obedience to the laws of thermodynamics. Divide the amount of energy available by the rate of output, and you get an upper probable limit of 100 million years for the Sun's total productive life. This is also, by implication, an upper limit to the age of the Earth as we know it. "As for the future, we may say, with equal certainty, that inhabitants of the Earth can not continue to enjoy the light and heat essential to their life for many million years longer unless sources now unknown to us are prepared in the great storehouse of creation." Kelvin wrote these words in 1862, and published them in a popular journal (*Macmillan's Magazine*).[11] In subsequent refinements of this calculation,[12] he would add further arguments, based for example on tidal friction and the dynamics of the Earth–Moon system, and in the light of fresh information about the thermal properties of rocks lower the range to some 20–40 million years, "and probably much nearer 20 than 40."

The impact was sensational. For by this time, as Kelvin well knew, a great deal was at stake. Darwin's *Origin of Species* had appeared just three years before the *Macmillan's Magazine* article. This had revolutionised our perspective on the world. It stated for the first time with complete clarity the modern view that species were not separately created but had evolved from simpler common ancestors by the operation of natural selection on the variations between individuals. The origin of these variations (what we now call mutations) was completely unknown, but it was clear that descent from a common ancestor must have been an extremely slow process, requiring what Darwin himself had described as "incomprehensibly vast... periods of time", with 20 to 40 million years much too little for all this to have occurred by natural selection. Nor did it help when Kelvin revised his 100 million year estimate of the age of the Earth sharply downwards, in the light of new evidence about the melting points of rocks. Indeed, Charles Darwin referred to Kelvin as an "odious spectre" and among his sorest troubles, and his son George was among the geologists most concerned with trying to find flaws in Kelvin's reasoning.

Radioactive Dating; the Gap Widens

The disagreement sharpened even further when Ernest Rutherford introduced radioactive dating. Rutherford, a New Zealander who did his research in Britain and Canada, was an outstanding experimentalist with little patience for high-sounding theory; he once remarked that he did not want to hear anyone talking about "the Universe" in *his* laboratory. Among other achievements, he discovered the difference between alpha, beta, and gamma radiation, and by using alpha particles as projectiles, found that nearly all the mass of an atom is concentrated in a tiny nucleus (see Chapter 2). He was, incidentally, firmly of the opinion that nuclear energy would never be a useable source of power.

Consider a grain of rock which contains a mineral that includes uranium. Over time, its radioactivity causes this uranium to decay, with helium and lead as the final products. We can estimate what fraction of the uranium decays each year, from the intensity of the element's radioactivity, and it turns out that the half-life (actual time for uranium-238 to decay to half its original amount) is 4.47 billion years. We can also estimate what fraction has decayed in a particular rock sample since it was formed, because the decay products will be locked in place where they were generated, along with the unchanged uranium. Putting these two bits of information together, we can work out an age for the rock.

For an igneous rock such as granite or basalt, this age is simply the time since the rock solidified. Sedimentary rocks are more complicated, but can be dated either from their relationship to igneous rocks (see Figure 1.1), or in some cases from chemical sediments formed within them. Rutherford applied this reasoning to a Cambrian rock in his possession, and reported it to be 540 million years old.

Rutherford initially analysed his rock samples for uranium and helium, but helium is a gas that can diffuse, perhaps even within rocks and certainly during normal sample preparation, so uranium/helium dating was quickly replaced by uranium/lead, still perhaps the most important single dating method for geologists. The uranium/lead method, as we now know, actually gives two estimates for the same sample, based on uranium-238/lead-206 (often written as $^{238}U/^{206}Pb$; the numbers next to the name of the element specify which isotope) and uranium-235/lead-207, providing an invaluable internal check. Different isotopes of an element have almost identical chemistry, but different nuclear properties and, where relevant, decay modes. Other important methods include potassium-40/argon-40 dating (half-life 1.25 billion years) for potassium-containing rocks, which are common; rubidium-87/strontium-87 dating (half-life 50 billion years) for extremely old objects, although the oldest objects that exist in our Universe are a mere 13.8 billion years old; and uranium-234/thorium-230 (half-life 245,000 years) for recently formed geological structures.[13] The half-lives themselves, as we now realise, are fixed by the laws of quantum mechanics and the fundamental constants of nature, which have been unchanged as far back in time as our telescopes can see, now around 12 billion years.

Rutherford's age for the Cambrian was much greater than even the geologists of his generation had thought. Indeed, for a while many geologists were reluctant to accept this great expansion of their timescale. After all, if the physicists had been wrong once in giving them too short a space of time, might they not be wrong again, but this time in the opposite direction? It took more than a decade, and major improvements in the techniques of radioactive dating and in their demonstrated mutual consistency, before these dates were finally accepted as definitive. Even now, problems and apparent inconsistencies resolved almost a century ago keep reemerging in the Creationist literature, like Transylvanian undead rising from their graves.

The currently accepted ages, based on large numbers of methods using these and other combinations of isotopes, are around 540 to 480 million years for Cambrian rocks, ~4.03 billion years for the oldest surviving rock formations on Earth, and 4.2 billion years for the craters of the Moon and for the oldest terrestrial rock grains so far discovered. The oldest known solid objects in the Solar System are calcium–aluminium rich inclusions in meteorites, and one of these has recently been dated to 4,568.2 million years ago.[14] This is in excellent agreement with 4.54 billion years for the Earth itself from lead isotope studies. There is even ongoing research into the time interval between the formation of the oldest meteorites, and that of the Earth–Moon system, using isotopic analysis of decay products from short lived radioisotopes present when the solar nebula was formed.[15] So at the time of writing, around 4.57 billion years ago is the best estimate for the start of the Earth's formation, but the process is thought to have taken up to 100 million years,[16] giving 4.47 billion years as the age of the Earth in its present form. Rocks from the first few hundred million years of the Earth's history have been lost either through tectonics or through bombardment; plausibly enough, since both of these would have been very active processes throughout that period; or, just possibly, they are still there to be discovered.

Although it is Rutherford's name that we associate with the introduction of radioactive dating, the first uranium/lead dating was actually carried out by his close associate, the American geochemist Bertram Boltwood, by then working at Yale.[17] Boltwood established that uranium is always accompanied by lead, that rocks from strata of comparable age have comparable lead/uranium (Pb/U) ratios, and that this ratio is greater in older rocks. He inferred that lead was the final decomposition product of uranium, and at Rutherford's suggestion calculated Pb/U ages by simply dividing the Pb/U weight ratio by the fraction of uranium that decays in one year, as measured by the rate of formation of helium from uranium-containing minerals.

Much more accurate Pb/U dates were collected by Arthur Holmes,[18] at the suggestion of Rutherford's rival Strutt, and read into the proceedings of the Royal Society in 1911, when Holmes was 21. (Strutt — later 4th Baron Rayleigh — was the son of the Rayleigh whom we meet again in Chapter 5 in connection with the discovery of the noble gases). In this paper, Holmes recalculated Boltwood's dates using the more accurate decay rate data that had become available, and added meticulous data of his own, derived from separate analyses of the different minerals in the same rock, a Devonian specimen from Norway. There are a couple of remarkable things about Holmes's paper. First is the fact that he is the sole author, although Strutt had suggested the problem and provided the mineral specimens used, and the laboratory space and equipment. Nowadays at least, Strutt would himself have been an author, and probably the senior author. Manners in science have changed, and not necessarily for the better. Secondly, the paper, like Boltwood's before it, actually contains an unnecessary approximation, and an instructive error. The approximation is that for some reason Holmes used a straight line equation for the amount of decay product formed over time, instead of the mathematically correct exponential decay (negative compound interest) law. He was fully aware that this was an approximation, and argued correctly that it would not make much difference. The error is that Holmes used the ratios of uranium to lead by weight, rather than by number of atoms. Since ^{238}U is some 15% heavier than ^{206}Pb, this means that (neglecting other distorting factors) all Holmes's calculated ages are 15%

greater than they should be. The missing material goes to make up the eight atoms of helium produced from each ^{238}U. These days, we would penalise a first-year undergraduate chemistry student for each of these shortcomings. Does this mean that a present-day student is expected to be the intellectual superior of Holmes or Boltwood? Of course not, but it is a salutory reminder of the intellectual difficulties faced by the first pioneers in this area, confronted with the shocking novelty of one element transforming itself into another. Holmes was active throughout his research life in refining these measurements, taking into account the existence of various isotopes of both uranium and lead, and finally came up with an age of 4.5 ± 0.1 billion years, based on the composition of modern igneous rock, in excellent agreement with the presently accepted value. Holmes was also noteworthy for his advocacy of continental drift (see below), a full generation before the discovery of plate tectonics led to its general acceptance by geologists.

Kelvin showed great scientific insight in his emphasis on gravity. Gravity is responsible for the birth of stars, driving the initial heating and compression of a gas cloud as it collapses inwards until it reaches the point where nuclear fusion can begin. It also determines the manner of their death, once their nuclear fuel is spent. Stars like our Sun will ultimately collapse to white dwarfs, poor relics of their former selves, while those much more massive will undergo catastrophic gravitational implosion, as we discuss in the final chapter of this book, spewing out newly produced elements as they form neutron stars and black holes. However, the one thing gravity cannot do is provide the steady supply of energy that the Sun and other stars emit throughout their lifetimes, given that these lifetimes must be as long as the radiometric dates require. There was clearly something wrong with both of Kelvin's arguments — what? What was driving Hutton and Lyell's apparently endless cycles of uplift and erosion? Are there any heroes or villains in this story, and if so, who?

Views Reconciled — Sources of the Missing Energy

Kelvin's arguments for the age of the Earth, and for the age of the Sun, both contain a similar assumption: that all the sources of energy have been taken into account. For the age of the Earth, he assumes that the energy is coming from the solidification of rocks. For the Sun, he correctly states that the most intense source of energy available to the science of his time is gravitational collapse, but he errs in assuming that the things he knows about are the only things that exist. We say this, however, with hindsight, and if in 1862 Kelvin had suggested that the facts required the existence of vast unknown sources of energy, he would have been justly criticised for the rashness of his speculations.[19]

What Kelvin could have no notion of was the enormous amount of energy available from nuclear processes. To take his two arguments in turn; the Earth is cooling much more slowly than he imagined. He made a very good job of estimating how fast the Earth is losing heat to space. However, he was mistaken in thinking that this heat was coming from the solidification of rocks as the Earth cooled. There is a major source that he did not know about continuously at work in the Earth, and that is natural radioactive decay. Through this process, the energy stored in the nucleus leads to the ejection of fast-moving alpha and beta particles, and the emission of gamma rays. All this energy is eventually converted to heat, and that amount of heat is substantial. Pierre and Marie Curie had observed that radium generates enough heat every hour to melt its own weight in ice, and while radium is, of course, much more intensely radioactive than the more abundant uranium that it is derived from, over geological time the amount of heat generated by the decay of uranium is enormous. We now know that only a very small fraction of the heat flowing up through the Earth's crust comes from rock solidifying. In other words, Kelvin was correct in saying that the heat flowing up through the crust represents energy being lost by the Earth to space, but incorrect in thinking that all, or even most, of this energy came from the latent heat of solidification of rocks. He made a good estimate of how fast the Earth was losing its reserves of heat, but did not know that these reserves of heat were being replenished.

There is a justly famous passage in Rutherford's correspondence where he speaks of addressing the Royal Institution in London — a world-class research centre, and a body devoted to the publicising of scientific knowledge — on the subject of radioactive dating, and seeing to his dismay that Kelvin, by then an old man, was in the audience. So, choosing his words very carefully, Rutherford improvised and said "Lord Kelvin had limited the age of the Earth *provided no new source [of heat] was discovered.*" Calling this a prophetic utterance, Rutherford described radioactivity as the new source that Kelvin had foreseen, and "Behold! The old boy beamed upon me."

We would like to imagine that Kelvin really heard in these words an echo of his words of over 40 years earlier about sources "prepared for us in the great storehouse of creation", but the facts argue otherwise. He continued to maintain that gravitation was the only possible source of the huge amounts of energy involved, and suggested that the energy of radioactive decay was secondary, having somehow been absorbed from other sources. However, few if any were convinced by this contorted reasoning, and soon physicists and geologists were attacking the question of exactly how much radioactive heating had contributed, and was continuing to contribute, to the Earth's energy balance.

Knowing what we now do, the importance of a "new source of heat" applies even more strongly to Kelvin's argument about the age of the Sun. The chemistry of the oldest known terrestrial rock grains shows that they were formed in the presence of liquid water. It follows that the Sun was already warming the Earth above freezing at least four billion years ago. We must therefore conclude that there is some source of energy to which the Sun has access, more intense than any that Kelvin knew about. That source is nuclear fusion, converting hydrogen to helium in the Sun's hot, enormously compressed core. The consequences of an energy-producing core were worked out by the English astrophysicist Arthur Stanley Eddington, who found an excellent agreement between calculations of heat flow in such a reactor and the brightness and temperature of different kinds of star. A star like the Sun is held in balance between gravitational energy pulling matter inwards, and the pressure of radiation streaming outwards, generated as gamma radiation in the star's core, and eventually getting converted to mainly visible light when it reaches the surface, which is at around 6,000°C and is glowing yellow hot. Right now, the Sun's core is converting hydrogen to helium. Eventually, it will run out of hydrogen. The process will stop, and the core will then collapse inwards, converting gravitational energy into heat, until it reaches temperatures and pressures high enough for the next step, conversion of helium to carbon. This process, "helium burning", needs roughly a fourfold higher temperature than "hydrogen burning". The electrical repulsion between two helium nuclei, each of charge +2, is four times as large as the repulsion between hydrogen nuclei of charge +1 (because the repulsion between two positive charges is proportional to both multiplied together), and this explains why helium burning only begins once the supply of hydrogen is exhausted. At this stage, the outer part of the star expands and becomes a *red giant*.

This fate, for the Sun, is estimated to lie some four billion years into the future, at which time the Sun will swell up and swallow Mercury, and in its later stages Venus and Earth. So the length of time left for life on this planet is roughly similar to the length of time since life first emerged here. Enjoy it while you can.

Stars considerably bigger than the Sun collapse more dramatically, giving rise to supernovae, and some small fraction of the energy of this collapse goes into making all the heavier elements. Just how this happens, will be discussed in the final chapter. The Earth, except for hydrogen and traces of primaeval helium, is made of recycled matter. We are stardust.

Plate Tectonics and Continental Uplift

It remains to account for the cycles of uplift and erosion. Erosion requires no special explanation, beyond the effects of *weathering* and gravity. There is no such obvious driving force for uplift, which was not really explained to the general satisfaction of geologists and geophysicists until around 1970.

For many years, it had been suggested that the continents moved relative to each other; Francis Bacon had noticed, for instance, that South America and Africa fit together like bits of a jigsaw puzzle. Moreover, there is a very good match between the fossil records, and also the rock types, in north-east South America and those in the bight of Africa, suggesting that these two continents had once been joined but had drifted apart over the past 60 million years. However, there seemed no way in which continents could move through the solid bedrock of the ocean floor.

A breakthrough here came with the surveys of the ocean floor itself, carried out in the 1960s. These led to the discovery of the Mid-Atlantic Ridge, where new crust is being created, and, by implication, the identification of such places as Japan and California as regions where crust is being destroyed. More extensive surveying showed that the entire crust of the Earth can be understood in terms of a dozen or so *plates*. In order to move relative to each other, the continents no longer need to move through the ocean floor bedrock. On the contrary, they ride on top of their plates, like luggage on a conveyor belt. The energy source for this process (known as *plate tectonics*) is convection in the mantle, as hot molten rock (*magma*) rises and cooler magma sinks. Molten rock lower down becomes hotter, expands, and rises to the top, while the less hot rock falls inwards because of its greater density. You can see the same kind of process at work in the rolling motion of water boiling on a stove, where the water at the bottom expands, becomes less dense than the layer on top, and rises, forming convection currents which drive along any scum floating on the water. In exactly the same way, the plates move as they are swept along by the convection currents in the mantle, and the ultimate energy source for these currents is the heating of the mantle by its own radioactivity.

Places like the Mid-Atlantic Ridge are known as *constructive* margins, margins because they are the boundaries between plates, and constructive because new crust is being formed. Here, the mantle is upwelling, the plates are being swept apart, and the new crust is generated from solidifying mantle material. Other places, such as the Western coast of the American continent, or the troughs of the Western Pacific, are known as *destructive* margins. At these, one plate is being forced beneath another, and crustal material is being returned to the mantle. Destructive margins are unstable regions, prone to earthquakes, and sculpted by volcanic activity. As the less dense part of the subducted crust forces its way back to the surface, it creates island arcs, such as Japan or the Philippines. For reasons that are not yet understood, from time to time the convection pattern changes, the plates change their directions of movement, and the continents begin a new dance. If any particular convection pattern lasts long enough, it is bound to lead to continents colliding at a destructive margin, and this collision leads to crumpling of the crust, which shows itself as mountain-building. We ourselves are living through the tail end of a massive period of mountain uplift, caused by the collision between the plates of India and Africa with that of Eurasia, forcing up the Himalayas and other mountain ranges across to the Alps and Pyrenees. The deep energy source of all these processes is the conversion of nuclear energy to heat by means of radioactive decay.

Lyell, you will recall, saw "no vestige of a beginning — no prospect of an end." How well has this summary stood up to a further 160 years of Earth studies? We see a very definite beginning, not on Earth (which was molten at its birth) but in meteorites and moon rocks. From the decay of radioactive isotopes, we can infer that the motion of the Earth's plates is now less violent than it was earlier; after all, four billion years ago, there was almost twice as much ^{238}U on Earth, more than twelve times as much ^{40}K, and something like 90 times as much ^{235}U. As for the end, we can now clearly foresee it, some four billion years in the future, from our knowledge of the fates of the stars. By then, the amount of radioactive material in the Earth's crust will have decreased even more, so that the cycle of uplift and erosion will be proceeding more slowly, but there is no reason to expect it to stop until the time when the planet itself is vaporised in the late red giant phase of our Sun.

To Sum Up

Rocks are laid down in strata, and processes of uplift and erosion can give rise to unconformities. Rocks must be older than any igneous intrusions that they contain.

The temperature increases as one descends into the Earth; therefore the Earth's centre is giving out heat. This heat arises mainly from the decay of radioactive atoms in the Earth.

Rocks (strictly, mineral grains in rocks) can be dated using radioactive isotopes and their decay products. The total amount of material (atoms of parent plus decay products) corresponds to the original amount present; we can compare this total with the amount of parent remaining, and from the ratio of these two numbers we can calculate how much time has passed since that mineral grain formed. Common dating isotopes are uranium-238/lead-206, and potassium-40/argon-40.

The Sun gets its energy from the fusion of hydrogen to helium. When it runs out of hydrogen, it will collapse inwards, until it starts converting helium to carbon. This is expected to happen in about 4 billion years time.

The continents move around on plates, which are driven by currents in the mantle, as the mantle gets rid of its heat by convection. It is the energy of this process, together with weathering, that drives the great geological cycles of uplift and erosion.

Some religious and philosophical traditions (Hindu, Buddhist, Aristotelian) regard the world as infinitely old and time as cyclical. Others (notably the Judaeo-Christian) regard it as having been created at some time in the past. Simple Biblical literalism implies a date around 4,000 BC, but allegorical counter-interpretations have been available for at least eight hundred years.

The modern science of geology dates to Hutton (1780s) and Lyell (1830s). They found evidence for cycles of erosion and uplift, with no evidence for any starting or stopping point, and assumed that the laws of nature and the actual processes operating on the Earth had been the same throughout time (uniformitarianism). By measuring the depth of sediments, 19th century geologists estimated the oldest rocks then known as dating back some 100 million years. However, they lacked an explanation of the driving energy behind geological processes.

Kelvin thought the Earth's heat came from the solidification of molten rock, and calculated from this assumption that it must have been a molten ball as recently as 40 million years ago. He also showed that gravity would give out much more energy than any chemical process for a body the size of the Sun, and estimated its age (from this, and its brightness) as 20 million years.

Rutherford in the early 20th century dated Cambrian rocks as around 500 million years old, in even greater conflict with Kelvin.

The problems were only resolved by taking into account energy sources that were unknown when Kelvin did his calculations; radioactivity slowing down the cooling of the Earth, and fusion as a long-lasting energy source for the Sun. The oldest presently known rock grains on Earth have been dated to over 4 billion years ago, while the study of meteorites gives an age for the Solar System of around 4.57 billion.

2 Atoms Old and New

What is now proved was once only imagin'd

William Blake

Really important ideas in science are not the work of a single individual or even a single generation. The idea of an atom, for instance, was developed by ancient Greek philosophers, revived by eighteenth century chemists to make sense of their discoveries about the composition of matter, and used by nineteenth century physicists to explain the effect of temperature and pressure on gases. Our modern idea of molecules, formed with definite shapes by joining atoms together according to definite rules, was developed by chemists studying naturally occurring substances in the late nineteenth century. In the early twentieth century, the structure of the atom itself was explained in terms of more fundamental particles, while the last half century has seen advances that make it possible for us to directly sense, and even move, individual atoms.

Atomic Theory in Ancient Times

Atomic theory dates back to the pre-Socratic philosophers, especially Leucippus and Democritus, who wrote and taught more than four hundred years BC. The works of the pre-Socratics survive only in fragments, and in quotations by later authors. For example, Epicurus, some 130 years later, built this theory into a unified view of the world and morality. The views of Epicurus were beautifully expressed by the Roman poet Lucretius, who lived at the same time as Julius Caesar, in his great work *De Rerum Natura* (On the Nature of the Universe). According to the ancient atomic theory, as we saw in Chapter 1, atoms are eternal and indestructible. All forms of matter are built up from a relatively small number of kinds of atoms.

What led the early atomists to theories so remote from simple appearances? Greek philosophers were greatly puzzled by the phenomena of change and motion. If something is real, how can it be transformed into something that it is not? If something is in one place, how can it move, since that would imply that it was no longer in that place? Besides, how can anything move without displacing something (if only the air) that is already there, in which case which one moves first? There are serious problems here, which were not properly solved until the mathematics of fluid flow and the theory of limits were developed in the 18th and 19th centuries.

One radical approach to the problems posed by change is to say that change itself is an illusion, and that the world of experience, in which we live and act, grow and die, is in some important sense unreal. Plato was influenced by this approach when he compared knowledge gained through the senses to a mere shadow-play on the walls of a cave. Such a view is deeply hostile to science, which relies on observation, and the influence of Plato and his followers was to greatly hinder the development of scientific thinking.

An alternative view, that taken by the atomists, is that Nature consists of two components; atoms, and the void. The atoms have room to move because there is a void between them. The individual atoms are unchanging, but their arrangement can change. All natural processes are caused by collisions and rearrangements of atoms. The force of the wind, for instance, is attributed to the impact of the "atoms" (we would now say "molecules") of air on objects. As Lucretius points out, the theory is consistent with common experience – for instance, the smell of a perfume diffuses through a room, so the perfume must be giving off invisible particles. Sheepskins hung up by the sea become wet overnight; this shows that "atoms" of water have found their way from the sea through the air to the sheepskin's surface. A pit dug a little inland from the sea fills up with drinkable water. This is because sea water contains more than one kind of atoms, and the jagged ones (which are also responsible for its sharp taste) stick to the ground through which the drinkable water flows (this is remarkably similar to the modern view, in which the *ions* that make up salt stick to clay). He could also have mentioned the way that things dissolve in water and can then be recovered unchanged when the water evaporates.

But let us allow Lucretius to speak for himself:

> *The headlong force of the wind lashes the body, overthrows great ships and scatters the clouds, then swiftly whirling strews the fields with tall trees ... So there must be unseen particles in the wind, sweeping the seas, the land, the clouds of the sky.*
>
> *Without seasonal storms, the Earth would not be able to give rise to the produce that gladdens us, nor could living things sustain themselves and propagate. So it is better to think that the same components occur in different things, much as the same elements occur in different words, rather than that anything could exist without an origin.*
>
> *Some things are primary objects, while others are assemblages, but there is no force that can destroy the primary objects, for in the end they overcome through their own sheer solidity.*[1]

In other words, rain, plants and animals must be made of similar atoms in different arrangements, in order to explain the facts of growth and feeding (Lucretius, of course, had no idea that the plants also needed carbon dioxide from the air, but otherwise his account is surprisingly modern). The force of the wind, and its ability to carry clouds, is due to the impulse of the "atoms" in the air. All the changes that we observe in matter are due to the movement of atoms and their combining together in new ways. For Lucretius, the atoms themselves are eternal and indestructible.

I should mention that this is part of a moral, as well as a natural, world view. Lucretius was writing at a time of great civil disorder. The constitution of the Roman republic, originally designed for a small city-state, was breaking down even as the territory under Roman control grew. The central authority was increasingly unable to control its own generals, and the civil liberties enjoyed by the old governing class, to which Lucretius belonged, were being trampled underfoot. At such times, individuals of spirit concentrate all the more keenly on the inner freedoms of thought and feeling. Thus Lucretius wrote to free the mind of superstitious fears through the light of understanding:

This fear and darkness of the mind cannot be dissipated by the rays of the Sun, nor by the clear shafts of day, but only by the perusal and understanding of Nature.

Why was there no progress from such a promising beginning? Bertrand Russell (History of Western Philosophy, Chapters IX, XXVII) suggests two possibilities, one deriving from the way society was structured, the other from political circumstances.

In the world of Greece and Rome, there was a total lack of contact between the useful arts (cookery and brewing, dyeing, metallurgy, tanning, ceramics, even medicine), which were delegated to the labouring classes, and the speculations of gentlemen. When labour is cheap, there is little pressure to improve technology, and when thinkers despise manual activity, they will not develop any experimental technique. Lucretius himself often appeals to observation, but experiment involves more than simply observing what presents itself. It is *the deliberate setting up of situations in order to observe them*, and no Greek or Roman gentleman would be likely to soil his hands in such a business.

Secondly, there was what Russell calls a failure of nerve. Leucippus and Democritus were great intellectual innovators; Epicurus and Lucretius were not. Leucippus and Democritus wrote as citizens of free states, which by the time of Epicurus had been subdued by Philip of Macedon and his son Alexander. Lucretius wrote in the last days of the Roman republic, when it was degenerating by way of civil war into a despotic empire. Later centuries saw the decline and fall of Rome, the chaos of Europe's Dark Ages, and the subsequent medieval reverence for authority and verbal argument rather than experience.

It seems relevant here that Lucretius's rationalist and naturalistic outlook did not endear him to the early Church. Thus St. Jerome, translator of the Bible into Latin, claimed, four centuries after the event, that Lucretius's death was suicide, that he had been driven insane by a love potion, and that his books, written in lucid intervals, had been reworked and corrected by Cicero. I have found no reason to believe that any of this is true.

Further south and east, in the Islamic and Hindu worlds, science and mathematics continued to flourish. Muslim theologians developed their own extreme version of atomic theory, in which matter consisted of dimensionless atoms, while space was atomised into separate points and time into instants, but the scientists of the Islamic world generally followed the continuous matter theories of Aristotle. There were major advances in the practical applications of chemistry, while the alchemists made important discoveries of how different substances react. However, a proper understanding of molecules, atoms, and the relationships between them requires an understanding of the mechanics of moving particles, and quantitative understanding of chemical reactions. We would have to wait until the 17th century for the first of these, and another hundred years for the second.

The Transition to Modern Thinking

"It seems probable to me, that God in the beginning formed matter in solid, massy, hard, impenetrable, movable particles... even so very hard, as never to wear or break in pieces; no ordinary power being able to divide what God Himself made one in the first creation." So wrote Sir Isaac Newton in his 1704 work, *Opticks*. Apart from the reference to God, there is nothing here with which Democritus would have disagreed. There is, however, very little that the present-day scientist would fully accept. In the rest of this chapter, we consider how atoms reemerged as fundamental particles, only to be shown, in their turn, as less than fundamental.

The Scientific Revolution and the Revival of Corpuscular Theory — 1600–1700

In 1543, on his death-bed, Nicolaus Copernicus received a copy of the first edition of his book, *On the Revolutions of the Heavenly Bodies*, in which he argued that the Sun, not the Earth, was the centre of what we now call the Solar System. In 1687, Isaac Newton published his *Mathematical Principles of Natural Philosophy*, commonly known as the *"Principia"*. With hindsight, and at the risk of oversimplifying, we can identify the period between these events as a watershed in the way that educated people thought about the world, and number the political revolutions in America and France, and the economic revolutions in agriculture and industry, among its consequences.

Before this *Scientific Revolution*, the prevailing view at least in Europe was still derived from that of Aristotle. It was assumed that the Earth lay at the centre (or the bottom) of the Universe. Objects on Earth moved according to their nature; light bodies, for instance, were thought to contain air or fire in their makeup, and these had a natural tendency to rise. Earth was corrupt and changeable; the heavens were perfect and immutable, and the heavenly bodies rode around the centre on spheres, because the sphere was the most perfect shape. By the end, Earth was one of several planets moving around the Sun; the movements of objects were the result of forces acting on them; the laws of nature were the same in the heavens as they were on Earth; and all objects tended to move in straight lines unless some force deflected them from this path. The Universe ran, quite literally, like clockwork. This mechanical world-view was to last in its essentials until the early 20th century, and still remains, for better or worse, what many non-scientists think of as *the* "scientific" viewpoint.

In 1611, Galileo turned the newly-invented telescope on the heavens, discovered sunspots, and moons round Jupiter, and realised that the belief in perfect and unchanging heavens was no longer sustainable. Earlier, he had studied the motion of falling bodies. In work that he started in 1666, Newton showed how the laws of falling bodies on Earth, and the movement of heavenly bodies in a Copernican solar system, could be combined into a single theory. To use present-day language, the Moon is in free fall around the Earth, pulled towards it by the same force of gravity that is responsible for the accelerating fall of an apple. This force gets weaker as we move away from Earth, according to the famous *inverse square law*, which says that if we double the distance, the force falls to a quarter of its value. Then with a certain amount of intellectual effort (involving, for example, the invention of calculus), Newton was able to work out, from the acceleration of falling bodies on Earth, and from the Earth–Moon distance, just how long it should take the Moon to go round the Earth, and came up with the right answer. He was also able to work out just how long it would take satellites at different distances to go through one complete orbit. Of course, at that time, Earth only had one satellite (the Moon), but six were known for the Sun (Mercury, Venus, Earth, Mars, Jupiter, Saturn), and his theory correctly predicted how the length of the year of these different planets would vary with their distance from the Sun (the answer is a 3/2 power law; a fourfold increase in distance gives an eightfold increase in time). Celestial and terrestrial mechanics were united.

It was around this time that a Dutchman, Anthony van Leeuwenhoek, began an extensive series of microscope studies, duplicating and greatly extending observations made by Robert Hook, whom we met in the last chapter. A superb experimentalist, he discovered the existence of hitherto unsuspected microscopic forms of life. Among the first to observe spermatozoa, he also described bacteria, yeast, the anatomy of the flea, and the cell structure of plants. He communicated his results to the Royal Society in London. Formally established around 1660, under the patronage of Charles II, this was and remains among the most prestigeful of learned societies. Here they caught the attention of Robert Boyle (of Boyle's Law for gases). Boyle tried to explain such properties of matter as heat, and the pressure of gases, in terms of the mechanics of small particles, or "corpuscles", and hoped that the other aspects of matter could be explained in the same kind of way. This was, after all, simply an extension downwards of the mechanical system that Newton had so successfully extended upwards. With hindsight, it is instructive to consider how far this hope was fulfilled. Atoms and molecules are in some ways similar in their behaviour to small objects obeying the everyday laws of mechanics, but in others they are very different, and, as we shall see in our discussion of quantum mechanics, it is these differences that must be invoked if we are to understand the forces involved in chemical bonding.

Early Modern Theory — 1780–1840

Between 1780 and 1840, chemistry underwent a revolution, that transformed it into the kind of science that we would recognise today. It is no accident that this was the same period as the beginning of the Industrial Revolution in Europe. Materials were being mined, and iron and steel produced and worked, on a larger scale than ever before. By the end of the period, mineral fertilisers were already in large scale use to feed the growing population. Demand for machinery led to improvements in engineering, and thus made possible improvements in the precision of scientific instruments. Much of the new interest in chemistry grew out of mining, mineralogy, and metallurgy, while improvements in manufacture and glass-blowing led to the precision balance, and to new apparatus for handling gases.

In the next two chapters, I describe this revolution in more detail. Here I just summarise some of the most important discoveries, as seen from our present point of view, and using today's language. This means running the risk of creating a misleading impression of smoothness and inevitability. Inevitability, perhaps yes; the world really is what it is, and once certain questions had been asked, it was inevitable that we would find the right answers. Smooth, no; the very concept of atoms, let alone bonding between atoms, remained controversial in some circles way into the 20th century. Outsiders sometimes criticise scientists for taking their theories too seriously, but at times they are all too slow to take them seriously enough.

Overall, mass is conserved; the mass of the *products* of a reaction is always the same as the mass of the *reactants*. This is because atoms are not created or destroyed in a chemical reaction.[2] Single substances can be *elements* or *compounds*, and the enormous number of known compounds can be formed by assembling together the atoms of a much smaller number of different elements.

The *combustion* of carbon (its reaction with oxygen) gives a gas, the same gas as is formed when limestone is heated. But there is no chemical process that gives carbon on its own, or oxygen on its own, by reaction between two other substances. Carbon and oxygen are *elements*, whereas the gas formed by burning carbon (what we now call carbon dioxide) is a *compound* of these two elements. The production of this same gas, together with a solid residue, by the heating of limestone, shows that limestone is a compound containing carbon, oxygen, and some other element.[3] To us, using today's knowledge, limestone is calcium carbonate, and *decomposes* on heating to give carbon dioxide and calcium oxide. In Lavoisier's time, there was no way of breaking down calcium oxide into simpler substances, and (as described in the next chapter) he therefore considered it to be an element.

Air is a mixture, and burning means reacting with one of its components, which we call oxygen. Metals in general become heavier when they burn in air. This is because they are removing oxygen from the air, and the *weight* (more strictly speaking, the *mass*) of the compound formed is equal to that of the original metal plus the weight of oxygen.

(Mass is an amount of matter; weight is the force of gravity acting on that matter. Atoms are weightless when moving freely in outer space, but not massless.) Different elements combine with different amounts of oxygen; these relative amounts are a matter for experiment. In modern language, when some typical metals (magnesium, aluminium, titanium, none of which were known when Lavoisier was developing his system) react with oxygen, they form oxides with the *formulas* MgO, Al_2O_3, TiO_2.

About one fifth of the air is oxygen, and if we burn anything in a restricted supply of air, the fire will go out when the oxygen has been used up. Nothing can burn in (or stay alive by breathing) the remaining air. Some materials, like wood and coal, appear to lose weight when they burn, but this is because they are in large measure converted to carbon dioxide and water vapour, which are gases, and we need to take the weight of these gases into account.

It was also shown during this period that the relative amounts of each element in a compound are fixed (*law of definite proportions*). For instance, water always contains 8 grams of oxygen for each gram of hydrogen. Moreover, when the same elements form more than one different compound, there is always a simple relationship between the amounts in these different compounds (*law of multiple proportions*). Hydrogen peroxide, also a compound of hydrogen and oxygen, contains 16 grams of oxygen for each gram of hydrogen. Similarly, the gas (carbon dioxide to us) formed by burning carbon in an ample supply of oxygen contains carbon and oxygen in the weight ratio 3:8, but when the supply of oxygen is restricted, another gas (carbon monoxide) is formed, in which the ratio is 3:4. Carbon monoxide is not intermediate in properties between carbon and carbon dioxide, but very different from both. For a start, it is very poisonous. It sticks to the oxygen-carrying molecules in the blood even more strongly than oxygen does itself, thus putting them out of action. It is formed when any carbon-containing fuel, not just carbon itself, burns in an inadequate supply of air. This is why car exhaust fumes are poisonous, and why it is so important to make sure that gas-burning appliances are properly vented. It is also one of the components of cigarette smoke, which helps explain why cigarettes cause heart disease and reduce fitness.

All these facts can be explained if the elements are combined in molecules that are made out of atoms, the atoms of each element all have the same *mass*,[4] and each compound has a constant composition in terms of its elements. For instance, each molecule of water contains two atoms of hydrogen and one of oxygen (hence the *formula* H_2O); hydrogen peroxide is H_2O_2; carbon dioxide is CO_2; carbon monoxide is CO; and the masses of atoms of hydrogen, oxygen, and carbon are in the ratio 1:16:12. Using these same ratios, we can also explain the relative amounts of the elements in more complicated molecules, such as those present in octane (a component of gasoline), C_8H_{18}, and sucrose (table sugar), $C_{12}H_{22}O_{11}$. Why C_8H_{18} and not C_4H_9, which would have the same atomic ratio? This can be inferred from the density of the vapour, using Avogadro's hypothesis (see below).

Thus, by the early 19th century, chemists were in the process of developing consistent sets of relative atomic weights (sometimes known as relative molar masses). However, there was more than one way of doing this; John Dalton (see next Chapter) thought, for instance, that water was HO and that the relative weight of hydrogen to oxygen was one to eight. This uncertainty, as we shall see in Chapter 4, led some of the most perceptive to question whether atoms were actual objects, or merely bookkeeping devices to describe the rules of chemical combination.

Evidence from the Behaviour of Gases (To Around 1860)

A French chemist, Joseph Gay-Lussac, noticed that the volumes of combining gases, and of their gaseous products, were in simple ratios to each other. In 1811, the Italian Count Amadeo Avogadro explained this by a daring hypothesis, that under the same conditions of temperature and pressure *equal volumes of gases contain equal numbers of molecules*. We now know this to be (very nearly) true, except at high pressures or low temperatures.

The background to this hypothesis, and why it took almost half a century for it to gain general acceptance, are discussed in Chapter 4. Briefly, it gives us a way of directly comparing the relative weights of different molecules, and of inferring the relative weights of different atoms.

For example, if we compare the weights of a litre of oxygen and a litre of hydrogen at the same temperature and pressure, we find that the oxygen gas weighs sixteen times as much as the hydrogen. (This is not a difficult experiment. All we need to do is to pump the air out of a one litre bulb, weigh it empty, and then re-weigh it full of each of the gases of interest in turn.) But Avogadro tells us that they contain equal numbers of molecules. It follows that *each molecule* of oxygen weighs sixteen times as much as each molecule of hydrogen.

One litre of hydrogen will react with one litre of chlorine to give two litres of hydrogen chloride gas. Thus the number of molecules of hydrogen chloride is twice the number of molecules of hydrogen, and twice the number of molecules of chlorine. So one molecule of hydrogen chloride contains half a molecule of hydrogen, and half a molecule of chlorine. It follows that the molecules of hydrogen and of chlorine are not fundamental entities, but are capable of being split in two. Making a distinction that is obvious in hindsight but caused great confusion at the time, each molecule of chlorine, or of hydrogen, must contain (at least) two separate atoms. By similar reasoning, since 2 litres of hydrogen react with 1 litre of oxygen to give 2 litres of steam, water must have the familiar formula H_2O, and not HO as Dalton had assumed.[5]

Avogadro's hypothesis was put forward in 1811, but it was not until 1860 or later that his view was generally accepted. Why were chemists so slow to accept his ideas? Probably because they could not fit it into their theories of *bonding*. We now recognise two main kinds of bonding that hold compounds together - *ionic bonding* and *covalent bonding*. Ionic bonding takes place between atoms of very unlike elements, such as sodium and chlorine, and was at least partly understood by the early 19th century, helped by the excellent work of Davy and Faraday in studying the effect of electric currents on dissolved or molten *salts* (see Chapter 9). They showed that sodium chloride contained electrically charged particles, and correctly reasoned that the bonding involved transfer of electrical charge (we would now say transfer of electrons) from one atom to another. But, as we have seen, Avogadro's hypothesis requires that many gases, hydrogen and chlorine for instance, each contain two atoms *of the same kind* per molecule, which raises the question of what holds them together. These are examples of what we now call *covalent* bonding or electron sharing, a phenomenon not properly understood until the advent of wave mechanics in the 1920s (see Chapters 11 and 12).

Physicists, meanwhile, were developing the *kinetic theory* of gases. This treats a gas as a collection of molecules flying about at random, bouncing off each other and off the walls of their container. This theory explains the pressure exerted by a gas against the walls of its container in terms of the impact of the gas molecules, and explains temperature as a measure of the disorganised *kinetic energy* of the molecules. The theory then considers that energy is distributed in the most probable (random) way among large numbers of such small colliding molecules. It can be shown that different kinds of molecule with different masses but at the same temperature will then end up on average with the same kinetic energy, and it is this energy that at a fundamental level *defines* the scale of temperature. This is a *statistical* theory, where, as discussed in Chapter 10, abandoning the attempt to follow any one specific molecule allows us to make predictions about the total assemblage.

The kinetic theory explains the laws (*Boyle's law, Charles's law*) describing how pressure changes with volume and temperature. Avogadro's hypothesis can also be shown to follow from this treatment. Many other physical properties of gases, such as viscosity (which is what causes air drag) and heat capacity (the amount of heat energy needed to increase temperature), are quantitatively explained by the kinetic theory, and by around 1850 the physicists at least were fully persuaded that molecules and, by implication, atoms, were real material objects.

Structural Chemistry, 1870 On

Chemists were on the whole harder to convince than the physicists, but were finally won over by the existence of *isomers*, chemical substances whose molecules contain the same number of atoms of each element, but are nonetheless different from each other, with different boiling points and chemical reactivities. This only made sense if the atoms were joined up to each other in different ways in these different substances. So atoms were real, as were molecules, and the bonding between the atoms in a molecule controlled its properties. This is what we still think today.

Einstein and Lucretius

The piece of evidence that finally convinced even the most skeptical scientists came from an unexpected direction — botany. In 1827, a Scottish botanist called Robert Brown had been looking at some grains of pollen suspended in water under the microscope, and noticed that they were bouncing around, although there was no obvious input of energy to make them do so. This effect, which is shown by any small enough particle, is still known as *Brownian motion*. Brown thought that the motion arose because the pollen grains were alive, but it was later discovered that dye particles moved around in the same way. The source of the motion remained a mystery until Albert Einstein explained it in 1905. (This was the same year that he developed the Special Theory of Relativity, and explained the action of light on matter in terms of *photons*). Any object floating in water is being hit from all sides by the water molecules. For a large object, the number of hits from different directions will average out, just as if you toss an honest penny a large number of different times the ratio of heads to tails will be very close to one. But if you toss a coin a few times only, there is a reasonable chance that heads (or tails) will predominate, and if you have a small enough particle there is a reasonable chance that it will be hit predominantly from one side rather than the other. Pollen grains are small enough to show this effect. But this is only possible if the molecules are real objects whose numbers can fluctuate; if they were just a book-keeping device for a truly continuous Universe, the effects in different directions would always exactly cancel out. And if molecules are real, then so are atoms. It is just as Lucretius said, looking at dust in the air two thousand years earlier:

> So think about the particles that can be seen moving to and fro in a sun-beam, for their disordered motions are a sign of underlying invisible movements of matter.

The Modern Era

Remember that for Newton, as for Democritus, atoms were eternal and unsplittable. The first subatomic particle to be discovered was the electron. Throughout the 19th century, physicists and chemists explored what happens when a voltage is placed across a gas. It turns out that if the pressure is reduced, current can flow, and the gas gives out light, as in the fluorescent light bulb. Even if all the gas is pumped out, some current still flows if the negative electrode (called the cathode) is heated. That is because it gives off *cathode rays* which carry the current. In 1896, the English physicist J. J. Thomson showed that these "rays" travelled in straight lines, could be deflected by electric or magnetic fields, and could make a screen coated with a suitable sensitive substance glow; *cathode-ray tubes* using these properties were used for televisions and computer monitors until the early years of the present century. Thomson's interpretation was that the cathode is emitting particles of negative electricity, which are fragments of atoms; these are what we now call electrons. By measuring how the beam of particles curved in a magnetic field, Thomson could calculate the ratio of electrical charge to mass for the electron.[6] He got the same answer, whatever metal was used for the cathode. Later, the same ratio was found for so-called beta-rays, emitted during radioactive decay, and it was correctly inferred that beta-rays are electrons, and that electrons are a component of atoms in general.

Since shortly before 1900, we have known that some atoms are unstable (radioactive) and break down from time to time, emitting high energy particles as they do so. These have so much energy that we can detect the effects of a single particle using a photographic plate or a *Geiger counter*. What we count with a Geiger counter are the death knells of individual atoms. The rates of radioactive decay, as we saw in Chapter 1, provide the clocks with which geological time is measured. The most spectacular kind of nuclear decay is neutron-induced fission, which lies at the basis of both civil and military uses of nuclear energy.

It was Rutherford (or, more exactly, Hans Geiger of Geiger counter fame, and Ernest Marsden, working under his direction) who first probed the more intimate structure of the atom. They bombarded a thin sheet of gold with alpha particles generated from the decay of radium, and found to their surprise that nearly all went pretty well straight through, with a very small fraction being deflected through large angles. It is from this work that we derive our picture of the atom as having over 99.9% of its mass concentrated in an extremely small *nucleus*, whose positive charge is balanced by the negative charge of electrons more spread out in space.

Around 1920, an instrument called the *mass spectrometer* was developed, using a combination of electrical and magnetic forces to find the mass[7] of individual atoms. Thus we can now weigh atoms one at a time. Doing this confirmed what had already been deduced from the behaviour of radioactive substances; many elements actually contain more than one different kind of atom, *isotopes* that have almost (not quite) identical chemical properties, but different masses. As discussed in Chapter 5, this explained the observation that very many elements have atomic weights close to a whole number, far more than could be expected by the operation of chance, but that some have atomic weights nowhere near a whole number. Each atomic nucleus is made up of protons and neutrons. Since each of these components has a mass very close to one, and the mass of the electron is very small in comparison, we can understand why each isotope has a mass very close indeed to a whole number.[8]

Studying high-energy collisions led to the conclusion that protons and neutrons were not simple objects, but each of them made up of three even smaller particles known as *quarks*. These are held together so tightly that the attempt to separate them is self-defeating; it requires so much energy that the attempt gives rise to new particles, bonded to the original quarks, so that such endeavours will for ever be frustrated. As to whether the quarks, themselves, have internal structure, or whether they are truly fundamental, that is perhaps one of the discoveries that will emerge from the Large Hadron Collider, now in the process of yielding its first results.

We will never be able to see the individual atoms in a piece of matter using an ordinary light microscope, because the wavelength of light is longer than the distance between individual atoms. But what we can do is to use x-rays (which are just like visible light but with very much shorter wavelength) to see *diffraction patterns* from crystals. This process, first demonstrated in 1912, gives a direct proof that the crystals really do consist of separate atoms spaced out in a regular manner. X-ray diffraction is now used routinely to find out just what the arrangement of atoms in a crystal is, and most of our knowledge of the structure of compounds has been attained in this way.

We can also use electrons themselves to "see" atoms using the *transmission electron microscope*. In this, high energy electrons are passed through thin films of matter. The electron microscope was invented around 1926, but it is only in the past few decades that it has been improved to the point that we can use it to image individual atoms.

In the past few decades, it has also become possible to detect individual atoms with probe-type microscopes. Such microscopes depend on having a tip of single atom sharpness. This is moved over the surface to be studied, controlling its position by using a piezoelectric crystal (very like the quartz crystal in a watch), which changes shape slightly when an electric field is applied to it. There are two common types of probe system. Force probes measure the force between the probe and the surface as the probe is moved over dimensions as small as a single atom at each step. Current probes measure electric current between the probe and the surface. It is often said that such microscopes make it possible for us to "see" individual atoms, but what happens is more akin to feeling and tasting than to seeing, despite the use of computer images. Using probes of this kind, it is even possible to pick up individual atoms and move them at will on the surface being studied. Should we call this chemistry, or the ultimate in small-scale engineering?

To Sum Up

The development of atomic theory can be divided into three main stages. The ancient theory was based on reasoning about the world, and on some facts of experience, but was developed without quantitative reasoning or any attempt at experiment. The period roughly from 1780 to 1890 saw the modern theory take shape, based on experiment, and on the observed quantitative regularities in the composition of matter and the behaviour of gases. The present stage (from 1890 on) uses sophisticated equipment to observe the collective and individual effects of molecules and atoms directly.

The ancient atomic theory was developed by the pre-Socratic philosophers in the 5th century BC, and elaborated by Epicurus (around 300 BC) and the Roman poet Lucretius (around 55 BC). This theory was based in part on common sense observations, like the force of wind and the facts of change and growth. It also had a moral agenda: to free people from superstition by showing that natural phenomena had natural explanations. However, there was no contact between these ideas and the technology of the time, and no attempt to deliberately set up tests of the theories. The disturbances of the times may also have helped lead to a failure of nerve, and a retreat from the search for knowledge. Moreover, the ideas of the atomists were overshadowed by those of Plato, who thought that the observable world was less real than the domain of thought.

The Scientific Revolution in the 16th and 17th centuries moved the centre of the Universe away from Earth (Copernicus), and succeeded in explaining the motions of terrestrial and heavenly bodies in terms of the same mechanical laws (Newton). These successes led to a belief that the same mechanical philosophy could be extended to the properties and changes of matter, and Robert Boyle sought to explain his discoveries about air pressure in terms of the mechanics of small particles.

The modern atomic theory dates from the late 18th – early 19th century, the time of the Industrial Revolution, when there was a great expansion of manufacture and a need to understand more closely what things were made of. From this time dates the distinction between atoms and molecules, the idea that there are some tens of different kinds of elements, the realisation that chemical reactions are changes in the way atoms are bonded together, and the notion of relative atomic and molecular masses. However, there was no agreement on a single scale of atomic masses, and much discussion as to whether atoms were real objects, or just book-keeping devices.

In the mid-19th century physicists developed the kinetic theory of gases, which depended on molecules being real objects in random motion, and this led to the acceptance, at last, of Avogadro's hypothesis that under the same conditions of temperature and pressure, different kinds of gases contained the same number of molecules per unit volume. The idea of bonding between unlike elements based on electrical charge had been accepted by chemists for many years; they now had to accept the idea of some other kind of bond, which could join similar atoms. Once they had done this, they were able to develop a theory of molecular shape, depending on the arrangement of, and angles between, such bonds.

In the 20th century, we discovered how to see individual radioactive decay events, how to infer the arrangement of atoms in crystals from their scattering of x-rays, how to "see" individual atoms using the electron microscope, and, most recently, how to "feel" or "taste" individual atoms using probe methods. Today's atoms are complex, highly structured entities, a long way from the eternal and unsplittable objects of the ancients. They consist of electrons surrounding a highly compact nucleus, this nucleus in turn containing protons and neutrons, and the protons and neutrons, once considered fundamental, are now regarded as being built up from quarks. There seems no reason at present to invoke even finer levels of structure, but this, of course, may change.

3 The Banker Who Lost His Head

If we had to name one person who, more than any other, turned physics into the kind of science that we can still recognise today, that person would have to be Isaac Newton. If we had to do the same for chemistry, we would choose Antoine Lavoisier. Newton, the first scientist ever to receive a knighthood, became president of the Royal Society, Member of Parliament, and died in his bed at age 84. Lavoisier became a senior tax official and financier (with special responsibility for tobacco), director of the Gunpowder Administration, spokesman for the Academy of Sciences, and at age 50 was guillotined for his services.

A Civil Servant Scientist

Lavoisier was born into a well-to-do bourgeois family in Paris in 1743. Socially ambitious in a society full of subtle gradations of status, he trained as a lawyer in order to qualify for the distinguished Order of Barristers of the Parisian *Parlement*, a traditional assembly with ill-defined consultative powers, and invested his inherited fortune in membership of the Company of Tax Farmers. This curious body was involved in the collection of indirect taxes throughout the whole of France, while its members individually lent money to the Crown, thus simultaneously taking on the highly respected roles of bankers, administrative civil servants, and investors in government securities.

At the same time, Lavoisier was busy advancing his standing as a scientist. He had been studying what we would now call physics, chemistry, and geology, while still a law student, and set about earning admission to the Academy of Sciences. This he achieved at the remarkably young age of 25, with a combination of pure science (composition of gypsum), and applications (problems of street lighting and water supply).

Lavoisier's administrative responsibilities made use of his scientific background, and came to include supervision of the tobacco trade, at that time a government-regulated monopoly. It was common practice among tobacco merchants to pad out their material with ash, for which Lavoisier was able to supply a simple test. Pour acid onto the adulterated material, and it will bubble and audibly fizz.

Much more important was his work with the Gunpowder Administration. Gunpowder is a mixture of saltpetre (potassium nitrate), sulphur, and charcoal. Before the discovery of South American guano deposits (see Chapter 6), this strategically vital material was obtained from fermenting organic matter with human and animal manure. Separating the saltpetre from manure was difficult, unpleasant, and a task intensely disliked, so much so that royal regulation of manure was among the causes of the English Civil War. French production had become haphazard, and a shortage of gunpowder was among the reasons for French defeat in the Seven Years War/French and Indian War of 1754–1763.[1] The Dutch had developed a system using beds of manure mixed with rotting vegetation, which Lavoisier copied, to such good effect that within a few years France was able to supply the material to its allies. French exports of saltpetre to the rebellious colonies played an essential role in providing this essential war material to the American colonists, and Lavoisier was able to write "One can truly say that North America owes its independence to French gunpowder".

Not for the last time in this book, we see science and technology intimately linked to warfare. The fruits of science inevitably increase our ability to do whatever it is that we want to do, and often what we want to do is kill foreigners.

Reform, Revolution, and the Terror

By the late 1780s, events in France had made some kind of reform inevitable. Foreign wars, culminating in French support for the colonists in the

American Revolutionary War, had left the country bankrupt, while bad harvests and economic slump in 1787–9 led to gathering social unrest. Various assemblies of notables were convened, and Lavoisier himself was prominent in the assembly of Orleans. Eventually, a meeting of the Estates General was announced for May 1789. This was the only body with the power to raise new taxes, but it had not met since 1614, and unresolved questions about its composition and powers were to prove critical. During this period, Lavoisier campaigned for moderate reform. He recognised the need for a strong executive, but regarded as unacceptable the fact that the King could pass laws by decree, even though he could only raise taxes by consent, since individual legal rights were surely at least as important as property rights. He also recognised that to be effective, the Estates General would have to be made more representative, to have defined legislative as well as fiscal powers, and to continue meeting at regular intervals. Reform, however, either constitutional or financial, was impossible in the face of the entrenched privileges of clergy, aristocracy, and the wealthier bourgeoisie of the parlements. When, in 1789, the Estates General finally met, the tensions within it set in motion the series of events that was to end in revolutionary Terror and, eventually, Napoleonic autocracy.

At its height, the revolution had no time for moderates, especially moderates who had been involved in the previous regime. Lavoisier, despite his brilliant success at the job, was removed from the Gunpowder Administration in 1792. The Academy of Sciences won a temporary respite by dominating the newly established Advisory Office on Practical Arts and Trades. Lavoisier himself worked within this Office, and was involved in developing the metric system, drafting a national educational curriculum, and advising on methods for printing banknotes; not a trivial problem before synthetic dyes became available. However, even before the revolution, allies of Marat had condemned the Academy as "elitist", and it was abolished by the Revolutionary Convention in 1793.

Wealthy bankers are rarely popular, especially during periods of economic unrest; tax collectors, never. Moreover, the members of the Company of Tax Farmers had exercised their functions in the service of the King, and so, once accused of financial irregularities, they didn't stand a chance. Lavoisier and other prominent members of the Company were arrested in November 1793, and imprisoned for over six months while their accounts were examined. When, at their trial in May, no evidence of fraud could be produced, the Revolutionary Convention simply declared them guilty of conspiracy against the people, and sent off Lavoisier with twenty-seven of his colleagues to the guillotine. Meantime, the armies of the fledgling Republic, using the products of Lavoisier's reformed Gunpowder Administration, were driving invaders from half a dozen nations back across their borders.

Three months later, the Convention had turned against itself and executed its extremist leadership, and the revolutionary Terror was over.[2]

Guinea Pig, Mesmerism, and Placebo

We start with two of Lavoisier's minor scientific accomplishments. "Minor" is a relative term, and either of these alone would ensure that we remember him.

Lavoisier used what was probably the original laboratory guinea pig in order to investigate what happens to food. How he did this illustrates his ingenious use of quantitative methods in general, and of the balance in particular. Lavoisier placed his guinea pig in the centre of a chamber surrounded by snow, weighed the amount of water melted by the guinea pig's body heat, and found that it matched the amount that could be melted by simply burning the amount of food consumed by the animal. In addition, he was well aware that the animal's metabolic activity required oxygen. He concluded, correctly, that respiration is, in effect, the combustion of food. This, of course, is the basic insight behind our own concept of the Calorie, strictly speaking a unit of heat, as a measure of food energy. We will have much more to say later about oxygen, combustion, and Lavoisier's definitive role in establishing a relationship between these.

One of Lavoisier's stranger civic duties was investigating the claims of one Dr Franz Anton Mesmer, who had abandoned his practice in Vienna after the scandalous failure of his treatments, and set up shop in Paris, to such good effect that two of his disciples had formed a joint stock company and franchise operation to exploit his techniques. The Academy of Sciences felt it necessary to investigate the phenomenon, and appointed a subcommittee that included Lavoisier, three of his colleagues, and a visiting American authority on electricity and related phenomena, by the name of Benjamin Franklin.

Mesmer claimed to produce his effects through the operation of what he called animal magnetism, which could be activated either by external magnets, or by the magnetism of another person, and his "mesmerising" hand gestures were said to be a method of producing such magnetic effects. The subcommittee would have none of it. They examined a group of patients, in whom Mesmer had actually produced fits (very therapeutic no doubt) by the use of magnets, and duly reproduced these phenomena in the same patients themselves. However, unbeknown to the patients, the magnets had been moved. The patients experienced the effects of the magnets where they believed the magnets to be, while their actual locations produced no effect whatsoever. Lavoisier concluded that the patients' experiences could be explained by the power of imagination, and that it was illogical in the circumstances to invoke the new phenomenon of an animal magnetic fluid. As we would put it today, Lavoisier was conducting a blind experiment,[3] and the cures that Mesmer claimed were the result of suggestion, or, in current terminology, the placebo effect.

None of this extinguished belief in animal magnetism.[4] Mary Baker Eddy, founder of Christian Science, believed that she was being poisoned by her enemies' manipulation of "malicious animal magnetism", while the US market in completely unproven magnetic curative devices has been estimated[5] at some $300 million annually.

Chemistry, Elements, and *The Elements of Chemistry*

We honour Lavoisier today, not for these peripheral intellectual adventures, but for a related set of ideas that lie at the heart of our modern understanding of the ways in which kinds of matter combine; the notion of a chemical element, the use of the balance, the implication that the amount of each element that goes into a reaction is equal to the amount to be found in the products, the correct identification of oxygen as an element, and our understanding of its role in combustion.

History has not been kind to Jean Rey, a French physician who in 1630 explained the observation that tin gained weight when heated in air and converted to its "calx" (oxide) by saying that the heavier particles of air stuck to its surface. He regarded this as a purely physical process, and is best remembered for an unfortunate comparison between the use of the balance and the use of reason; "The latter is only employed by the judicious; the former can be practised by the veriest clown." The balance told him that the tin was removing something from the air, but his reason told him to misinterpret this crucial observation.

The balance had been used in metallurgy since antiquity, both to weigh objects and to determine their purity by finding their density (weight divided by volume). We have the famous legend, written 200 years after his death, of Archimedes puzzling about how to find the volume of a crown made by a jeweller for King Hiero of Syracuse, seeing how the water in his bath rose as he stepped into it, and running down the street stark-naked shouting "Eureka! I've got it!" when he realised that all he needed to do was put the crown in a measuring cylinder full of water, and see how much the water level rose.

What was new about Lavoisier's use of the balance was the idea of comparing the weights of the starting materials with those of the final products. When he started his career, many reputable scientists sincerely believed that water, given enough time, would eventually turn to stone, and pointed to the residues left behind, when water was evaporated after prolonged boiling, as evidence for this process. Lavoisier, remember, was concerned with the purity of water in connection with the practical problem of drinking supplies, and looked into this question at the very outset of his scientific career. He kept water boiling in a device called a pelican (essentially, a piece of apparatus in which water was boiled, and the steam condensed so that it ran back into the vessel) for over 100 days. He then poured off the water, evaporated it to dryness, and found that there was indeed a solid residue, which he weighed. So far, nothing he had done was very different from what others had done before him. But remember that Lavoisier was a banker, much concerned with seeing that what came out of the balance sheet was the same as what went in. So before he started, he had weighed the pelican, and when he weighed it again at the finish, he found that it had lost weight, equal in amount to that of the solid residue recovered from the water. From this, he correctly inferred that the water had not changed its nature at all, and that the solid residue was merely material that had dissolved over time in the water from its container. What goes in is the same as what comes out,[6] what we now call the principle of conservation of mass, or, when applied to a single process, the determination of mass balance.

Concern with mass balance quickly led Lavoisier to note that combustion involved removing material from the atmosphere. It was common practice at that time for scientists to establish their priority by publishing encrypted versions of their discoveries, or by depositing sealed accounts. (Now, of course, priority goes to whoever published first, a much healthier situation.) In 1772, Lavoisier deposited a sealed account with the Academy of Sciences, in which he noted that sulphur and phosphorus, on burning, gained considerably in weight, implying that they had extracted material from copious amounts of air. This, of course, in contrast with the burning of organic matter, in which material (attributed to a mysterious substance called phlogiston) was apparently lost.

In a way, there was nothing very novel in this result. As we have seen, it had already been known for over a century that the changes metals underwent when heated in air, converting them to a "calx" or chalky, powdery substance, involved an increase in weight. With hindsight, it is obvious to us what is happening, but there were several obstacles to understanding at the time. The idea that air might contain more than one substance, instead of being the element assumed from antiquity, was still new, although Joseph Black in Glasgow had shown that heating limestone led to its giving off what he called "fixed air", a gas that could extinguish flame and, in a reversal of the reaction used to produce it, turn lime water cloudy by forming finely divided chalk.

The next step depending on correctly interpreting an observation that had already been made by Joseph Priestley in England; that while mercury is converted to its calx by heating in air, stronger heating of that calx would decompose it to the original metal, and to a gas which could support respiration longer than ordinary air. Lavoisier himself found that heating the calx with carbon gave rise to "fixed air", which he identified by its ability to extinguish flame and to give a solid precipitate with lime water, but this was different from the "respirable air" produced by heating the calx on its own. Furthermore, Lavoisier showed that formation of the calx involved uptake of "the most respirable part" of ordinary air. This was part of a more general observation, which Lavoisier recorded as early as 1772, that conversion of a metal to its calx involved the uptake of what he described as an acidic substance present in the air.

With hindsight, all that remained was to join up the dots, and joining up the dots was inevitable. As to why it was Lavoisier who took the crucial step, it is worth recalling that he had been trained as a lawyer, a profession involving the precise use of unambiguous language. It is also worth noting the complaints he made about the way in which he himself had been taught chemistry. He contrasted the teaching of mathematics, where each step is shown to follow from the preceding, with that of chemistry where the standard presentations "make use of terms that have not been defined, and suppose the science to be understood by the very persons they are only beginning to teach."[7]

For this reason, Lavoisier deliberately set about reforming the *language* of chemistry, an ambitious task that Thomas Jefferson, who knew of his project, regarded as premature. In the way of these things, what was originally intended as the formal record of a memoir on the subject, that Lavoisier read to the Academy of Sciences in 1787, expanded into a full-scale textbook, *The Elements of Chemistry*, published just two years later (here again we can only admire Lavoisier's energy).

If you want to reform the language of a subject, the obvious place to start is with the definition of its key concepts, and the key concept of chemistry, then as now, was that of an *element*. For the ancients, as Lavoisier pointed out, there had been the four elements of earth, air, fire, and water. To these, the alchemists had added salt and sulphur (he could also have mentioned mercury), while others had spoken of different kinds of "earth". Faced with this, Lavoisier concluded that instead of attempting to find the nature of "those simple and indivisible atoms of which matter is composed", we should define an element as a substance that cannot be broken down into simpler components. This provides what we would now call an "operational definition"; an element is defined in terms of experimental tests. It also — as Lavoisier explicitly recognised — took account of the possibility of changing knowledge. Thus air had already been shown to contain more than one component, so it was not an element, and water also was shortly to suffer a demotion.

Things would now fall into place very quickly. "Respirable air" could not be broken down any further, nor could it be made from simpler components, so it qualified as an element. This Lavoisier called *oxygen*, producer of acidity, since from his work with phosphorus and sulphur he believed it to be a component of all acids. Mercury was an element. So was carbon, the purest form of charcoal. Carbon reacted with oxygen to give "fixed air", which was therefore not an element but a compound. In the reaction between charcoal and the calx of mercury, carbon was decomposing the metal oxide, leaving the metal free. A much more important reaction of the same type had been practiced, although not understood, for over 6,000 years, ever since the discovery (Chapter 9) of how to extract metals from their ores. Calxes were compounds between the various metals and oxygen, while the metals themselves were elements. So on moderate heating, mercury reacted with oxygen from the air to form an oxide, but stronger heating would decompose this oxide to its original elements. For an understanding of why the reaction would reverse itself in this way, we would have to wait for the science of *thermodynamics*, discussed in Chapter 6. Once oxygen had been extracted from the air, what remained was an inert residue, which Lavoisier named "azote", or lifeless. This is still what it is called in French, although the modern English name is *nitrogen*, in recognition of its presence in nitre.

What about water? Lavoisier knew about "inflammable air", a gas of low density formed when metals react with acids, and was indeed the first to suggest its use in balloons. Believing that its reaction with oxygen would produce another acid, Lavoisier in combination with Laplace (now best remembered as a mathematician; see Chapter 10) allowed oxygen and inflammable air from pressurised tanks to react within a glass bulb, and collected the liquid formed. To their surprise, it was nothing more nor less than pure water. Hence the name *hydrogen*, producer of water, for inflammable air. They were not the first to carry out this experiment, and to verify that the weight of water was equal to the combined weight of the reacting gases, but they were the first to recognise its significance. If water could be produced from two simpler substances, then whatever Aristotle may have said on the subject, it was not itself an element.

Not content with determining the nature of water by *synthesis* (combining elements together), Lavoisier verified his result by *analysis*. Again, he was not the first to carry out the crucial experiments, but the first to understand them. He reacted iron with water, collected the gas produced, and verified that this was indeed hydrogen by burning it and recovering the original material.

We can summarise the "new chemistry" that Lavoisier developed throughout the 1770s and 1780s as follows:

1) Combustion or calcination involves the formation of a compound with oxygen (the obsession with phlogiston had prevented Priestley, who had isolated oxygen, and Cavendish, who had independently produced water by burning hydrogen, from realising this).
2) Air is a mixture of oxygen and azote.
3) Water is a compound of hydrogen and oxygen.
4) Caloric or heat substance is an element that is released during combustion.
5) Combination with oxygen produces acids, and all acids contain oxygen.
6) Matter is composed of simple bodies (i.e. elements) free or in combination. Bodies are regarded as simple as long as all efforts to simplify them further or to produce them by combining other bodies have failed.
7) Matter is conserved during chemical changes, as is the amount of each of the "material principles" i.e. elements.
8) A new notation was called for to express these facts.

In The Rearview Mirror

It's worth a look at how well these conclusions have stood up over more than two centuries. (1) and (3) read as true today as they did in 1789, when Lavoisier collected his ideas in his *Elements of Chemistry*. (2) is something like 99% true. In addition to nitrogen and oxygen, air contains about 1% argon (see Chapter 5) and small but vitally important minor amounts of carbon dioxide (see Chapter 13), as well as traces of other gases. (4) is an idea that was already close to outliving its usefulness. In 1798 Count Rumford showed conclusively that heat could be generated from friction, and concluded correctly that it was a form of disorganised motion, although the concept of caloric was to survive for a few decades longer and lay behind John Dalton's thinking.

(5) seems a little bit strange, even from the perspective of the time. As Lavoisier himself helped establish, the reaction of metals with air is combination with oxygen, and metal oxides are generally speaking basic, rather than acidic. We should, however, remember that he used the terms

"acid" and "base" slightly differently from the way that we do. Thus he speaks of sulphur as being the "base" that combines with oxygen to make sulphuric acid.

Many acids do in fact contain oxygen, and it is often true that the more oxygen a compound contains, the greater its acidity. To take an example whose essential features were familiar to Lavoisier, sulphuric acid (H_2SO_4) is a much stronger acid than sulphurous acid (H_2SO_3). However, the "muriatic acid" released when sulphuric acid is poured onto common salt turns out to contain no oxygen at all. So we can regard (5) simply as an empirical generalisation, which seemed reasonable at the time, but which is now rejected in the face of fresh evidence. That's how science works.

(6) is essentially what we think today, with a "simple body" being the same thing as what we call an element. Lavoisier himself listed 33 of these, of which 23 are still recognised as such.

It is worth looking at those that haven't made the cut. Lavoisier starts his table with light, followed by "caloric", which he identifies with heat or fire. For him, and for his contemporaries, this was a kind of fluid, and expansion and contraction were the results of having more or less of this fluid in between what he referred to as "the molecules of bodies". He also refers to the "radicals", or so far unidentified component parts, of what we now recognise as hydrochloric, hydrofluoric, and boric acids. Here his information was not so much inaccurate as incomplete. Finally, he lists five materials that are in fact compounds, but could not have been further analysed by the methods of his day; lime, magnesia, barites, argil, and silex. These we now recognise as calcium oxide, magnesium oxide, barium sulphate, aluminosilicate clay (a whole rich family of compounds in itself), and silica sand.

One idea that Lavoisier seems unconsciously to have inherited from the alchemists was that elements could directly convey properties or principles to the compounds that contain them. We have already seen that he thought oxygen conveyed the principle of acidity, and he also believed that it imparted redness. After all, lead, mercury, and iron all have red oxides, and blood is reddest when most fully oxygenated.

(7) is the law of conservation of mass, and was to provide the vital link between experimental chemistry and the atomic theory of the ancient philosophers.

As we have seen, Thomas Jefferson regarded (8) as a premature exercise. He was wrong, and much of Lavoisier's nomenclature is still with us. Lavoisier spoke of –ous and –ic acids, depending on their oxygen content, much as we do today, and he also spoke of metal oxides, and identified salts by reference to the metal (he would have said the base) and the acid that they contained. Thus what we call potassium nitrate is what he called nitrate of potash. Lavoisier's stated aim was to establish an unambiguous nomenclature in which the names of substances were systematically related to their composition, as an aid to clarity of thought, and in this he was wholly successful.

If he had survived the Terror, Lavoisier would quite probably have lived another 20 years or more, long enough to witness and contribute to the next round of developments. His clarity of language would at the very least have helped make clear the underlying issues, and might well have saved the world of science half a century of unnecessary confusion.

To Sum Up

Lavoisier was an administrator, a civil servant, a banker, and an outstandingly clear-headed and insightful experimental scientist. His very real services to France under the *Ancien Régime* led to his execution during the revolutionary Terror, a mere three months before its ending. Lavoisier transformed chemistry into the science we know today, giving their modern meanings to the ideas of element, reaction, conservation of mass, and combustion, and recognising oxygen and hydrogen (among others) as elements, water as a compound between them, and air as a mixture of oxygen and nitrogen.

Had he survived the Revolution, his insights might well have shortened the period of confusion described in the next chapter by decades.

4 From Particles to Molecules, with A Note On Homoeopathy

Lavoisier, to whom we owe our modern concept of an element, sought the limelight, promoted his own reputation, rose from bourgeois origins to wealth and prominence, was at the centre of Parisian public affairs, and was executed for political reasons when he was still at the height of his powers. John Dalton, to whom we owe our modern concept of an atom, affected an almost excessive modesty, spent most of his adult life as a provincial school teacher, rarely visited London, paid scant attention to the honours it had to offer him, died of natural causes at age 78, and was buried with full honours. Lavoisier was a victim of the French Revolution. Dalton was part of the community of innovators and thinkers who gave birth to the Industrial Revolution. Lavoisier gave as clear a definition of an element as was possible at the time. Dalton's concept of an atom contained built-in errors that would hamper chemistry for a generation, until science was ready to accept an idea that Dalton himself had rejected, but was later articulated by a neglected (one might even say justly neglected) Italian contemporary.[1]

Obscure but Famous

John Dalton was born in 1766 into a Quaker family in the village of Eaglesfield, on the fringe of the English Lake District, where by age 12 he was fulfilling the role of schoolmaster. A few years later, he and his brother were running a school together in Kendal, and he used his leisure time to study Latin, Greek, French, and mathematics with what was then called natural philosophy. In later life, he perhaps overplayed these humble origins, giving insufficient credit to the supportive network of the Society of Friends, and to the books that he had access to at this stage. Dissatisfied with the poor pay of a schoolmaster, he contemplated taking up a profession. At this stage the only possibilities seem to be law and medicine, and the financial and institutional barriers to both were insurmountable, especially as Nonconformists were not admitted to either of England's ancient universities. In retrospect, we would describe him as part of the first generation of a new profession that he helped create, and which in 1834 received the name by which we still call it, *scientist.*

Dalton's earliest research was in meteorology, a subject in which he retained his interest for the rest of his life. He also, when considering a career in medicine, conducted a curious series of experiments in which he weighed his food and drink, and his excreta, in order to infer how much weight he was losing in perspiration. In 1793 he moved to Manchester, where apart from occasional excursions he lived until his death in 1844, and was appointed teacher of mathematics and natural philosophy at the newly opened Manchester New College. (This was the successor to the Warrington Academy, where Priestley, a Unitarian and discoverer of oxygen, had taught.) In 1797, having attended a course of 30 lectures on chemistry by a visitor, one Doctor Garnett from Harrogate, he added chemistry to the subjects he taught there.

In 1800, he decided to go freelance, and opened his own "Mathematical Academy". Here he soon had eight or nine pupils at 10 guineas a year, plus upward of 20 lessons per week, privately, at two shillings each. Not affluence, but enough, and an income that he was to supplement with lectures at various universities and institutions. In 1833, in recognition of his services to science, he was awarded a government pension of £150, later increased to £300. Of his students, only one, James Prescott Joule, became himself a notable scientist.

Dalton's personal life is perhaps best described as uneventful. He never married, and stayed in various lodgings, living alone for the last 14 years of his life. He would light the fire in his rooms in the Literary and Philosophical Society buildings, return home for breakfast, spend the morning teaching and the afternoon and evening in the laboratory. A game of bowls every week, with his green fees duly noted in his account books. He suffered a minor stroke in 1837, two more serious strokes in 1838 and 1844, and in July 1844 was found dead in his bedroom by his attendant.

It would be a mistake to think of Dalton as exclusively a "pure scientist". His range of interests reflected the concerns of the Industrial Revolution, in which the city of Manchester played such an important role. He was interested in the behaviour of water vapour because of the importance of the steam engine, among other reasons. Gaseous hydrocarbons were important in connection with street lighting, and Dalton acted as consultant to Manchester's gasworks, and gave evidence to Parliament about gas making. Lancashire was emerging as the world's leading centre of cotton manufacture, making the analysis of dyestuffs a subject of great importance, and this too Dalton was involved in. Like Lavoisier, he was interested in the analysis of drinking water for the growing cities, and Manchester itself was expanding dramatically throughout his time there. There was a direct connection between his studies of the solubility of gases — of which much more will be said in the following pages — and his friend William Henry's commercial interest in the production of aerated (i.e. carbonated) water.

In 1794, Dalton was elected to membership of the Manchester Literary and Philosophical Society, which functioned as a club where scientists and engineers from the area met and exchanged ideas. This body, founded in 1781, was of more than local importance, and its *Memoirs and Proceedings* were for a while the only scientific journal in the UK other than the *Philosophical Transactions* published (still published) by the Royal Society. Its honorary members at various times included Lavoisier, Volta, Benjamin Franklin, Erasmus Darwin (Charles's grandfather, and an early advocate of the doctrine of evolution), and Priestley. In 1802 he was elected to the committee of the Literary and Philosophical Society, and became vice president in 1808, and president from 1817 until his death. He gave numerous and moderately remunerative public lecture courses, at the Royal Institution in 1803, and at numerous locations from 1805 on. Such public lectures depend in large part on demonstrations, and Dalton, in accord with the entrepreneurial spirit of his time and place, built up his own stock of equipment, and his personal library. He paid little attention to London's Royal Society, of which he became a Fellow only in 1822 after friends had proposed him without his knowledge, but was very happy at his 1816 election to the French Academy of Sciences. In 1831 he was involved in the foundation of British Association for the Advancement of Science, which later became a major intellectual forum, and was elected vice president in 1836. When he died in 1844, Manchester awarded him a civic funeral, and more than 40,000 people filed past his coffin.

Dalton was red–green colourblind, and his first major lecture to the Literary and Philosophical Society was on this subject, whose symptoms he described meticulously. A few years later, Thomas Young correctly described human colour vision in terms of three different kinds of colour-specific receptors, and connected this to the wave theory of light. Dalton did not accept this theory, and thought his vision defect was due to his eyeball absorbing red light. He gave orders for his eyeballs to be tested for this after his death, when they were found to be normal, in agreement with Young's theory.

The Development of Dalton's Thinking

In 1793, shortly after moving to Manchester, Dalton met William Henry and formed a friendship that was to last until Henry's early death in 1836. Dalton had designated William Henry as his biographer and custodian of his papers, a duty that was passed on by default to his son W.C. Henry, who did not however take this obligation sufficiently seriously. For this reason, and because many of Dalton's papers were destroyed in a fire started by an air raid in 1940, we know very much less about the evolution of his ideas than we would like to.

Nonetheless, we have enough of his writings to trace his intellectual development, and we can see that it does not fit into any of the stereo-types favoured by philosophers of science, still less by those who believe in the damaging fiction of a single "scientific method". Dalton did not observe regularities in combining weights and infer the existence of atoms, as required by the "inductivist" theory of science. Still less did he start from some fully formed theoretical insight, work out the implica-tions, and satisfy himself that they were true, as required by the rival "hypothetico-deductive" theory. In reality, he was led to his insights step-by-step, by way of misleading models and muddled analogies. I think this is the way that really important ideas usually get started. There are recipes enough for testing such ideas, but no recipes for generating them in the first place.

The next few paragraphs attempt to recapture Dalton's thinking at this point. They are among the most difficult in this book, because they require us to see the world through Dalton's eyes, and that means imagining that we don't know many things that are actually very familiar to us. The reader who wishes may simply skip to the next section, but will miss out on an interesting and surprising detective story.

Dalton's first and longest lasting interest was meteorology, the subject of his first book in 1793. This did not attract much attention at the time, but contained the first definition of the "dewpoint", the temperature at which water will condense out of the air. Perhaps we see here the effects of Dalton's rural upbringing. The discussion of dew also included the novel suggestion that evaporation and condensation of vapour are not the effects of chemical affinities, but that "aqueous vapour is a fluid *sui generis* [of its own kind], diffused among the rest of the aerial fluids." In other words, the water vapour present in damp air is not chemically combined with the other substances present, but is acting independently. Water vapour can press down on water vapour and make it condense, but the other gases have no such effect.

By 1801, Dalton had generalised this idea of independently acting fluids to describe the behaviour of gases in general. To understand his achievement here, we need to think back to the theories of his time regarding gas pressure. In the 17th century, Robert Boyle (of Boyle's Law fame) thought of gases as being made up of springlike particles that pushed against each other, and it was this mutual repulsion that was responsible for the pressure any gas exerts against the walls of its container. Dalton had a rather similar theory. He thought of each of the gas particles as being surrounded with a certain amount of "caloric" or heat, which resisted compression. So a gas was rather like a collection of small soft round sponges that resist being pressed more closely together. If you heat up a gas at constant pressure, then according to Charles's Law (already known to Lavoisier, but independently observed by Dalton), different gases expand by the same amount. This fits Dalton's model if the different gases are absorbing equal amounts of caloric. Our present explanation, that the pressure is actually due to the impact of particles bouncing off the walls, was still some decades in the future.

This presented Dalton with an interesting problem. Air, as Lavoisier and others had already established, is roughly four fifths nitrogen and one fifth oxygen. If you mix any volume of oxygen with four times its volume of nitrogen, all at ordinary atmospheric pressure, you end up with a total of five volumes of air at the same pressure. This seems obvious enough, but there are times when the mark of deeper understanding is to be puzzled by the obvious, and this is one such time. According to Boyle's Law, nitrogen expanded from four volumes to five volumes, so it should be exerting 4/5 the ordinary atmospheric pressure.[2] Likewise, oxygen has expanded from one volume to 5 volumes, and is exerting 1/5 its original pressure. So the total pressure (one atmosphere) is the sum of these two contributions, and *each gas is exerting its pressure independently of the others*. This is still known as "Dalton's law of partial pressures", and immediately established his international reputation.

Isn't this all rather obvious? To us, yes; to Dalton and his contemporaries, no. In Dalton's model, particles of nitrogen can push against particles of nitrogen, and particles of oxygen can push against other particles of oxygen, but the particles of oxygen and nitrogen do not push against each other. Why not?

So Dalton had already been thinking about how water (as vapour) mixes with air, and how different gases in the air are mixed with each other. The next thing to be considered was how different gases mix (or dissolve) in liquid water. This solubility is not very large, although the dissolution of oxygen in water is obviously of vital importance to aquatic life. It was around this time (in 1802 in fact), that William Henry formulated what is still known as "Henry's Law", which says that the solubility of any gas in water is directly proportional to its pressure; such information, as we have seen, was of practical as well as theoretical interest to a manufacturer of aerated water. Dalton took the logical step of considering what happens when you dissolve a mixture of gases, such as air, and was able to point out to Henry that each individual gas dissolved according to its own partial pressure. Once again, the particles of the different gases are not interacting with, or competing with, each other.

So far so good, but what exactly is happening when gases dissolve in water at all? And why do different gases have different solubilities? Dalton imagined that the particles of gas were fitting into pores in between the particles of the liquid. (He saw water itself, of course, as composed of particles, because this explained how it could exist as vapour. Lucretius would have thoroughly approved.) If so, different kinds of gas must fit into different kinds of pore. For if not, all gases would have the same solubility, and the presence of one gas, by blocking off pores, would reduce the solubility of another. This means that *different gases are composed of different kinds of component particle, with different properties*. And if this is true of gases, it is going to be true of everything else as well, since Dalton realised that, just like water, substances in general could be prepared as solid, liquid, or gas, depending on temperature and pressure.

We do not have Dalton's notes for the lectures he gave at the Royal Institution in the winter of 1803/1804, but according to a reviewer he referred explicitly to a theory of *atoms*. Certainly, by 1805 he was explaining the law of partial pressures in terms of different sizes *and weights* of different kinds of component. Building on Lavoisier's foundation, he went on to explain chemical combination in terms of the attractions between these different kinds of component. In 1807, he presented his ideas in a series of lectures at the universities of Edinburgh and Glasgow, saying, "I chose Edinburgh and Glasgow in preference to any other cities in Britain because I apprehend the doctrines inculcated within those cities meet with the most rigid scrutiny, which is what I desire." The syllabus for the lectures included "elastic fluids conceived to consist of indivisible particles or atoms of matter, surrounded with atmospheres of heat", "reasons for believing that in the chemical union of elementary principles, we generally, if not always, find a compound consisting of one atom of each," and "coincidences of these with experimental results". Here we have fully established the modern notion of elements consisting of fundamental particles, or atoms, of these atoms as real material things, and of different kinds of element being composed of different kinds of atom.

Dalton's Atomic Theory

In 1808, Dalton summarised his thinking in his *New System of Chemical Philosophy*. In this he states that the number of ultimate particles is unchanging, and speaks of the significance of "the relative weights of the ultimate particles, both of simple and compound bodies, the number of simple elementary particles which constitute one compound particle, and the number of less compound particles which enter into the formation of one more compound particle." He used his own system of symbols for the elements, spheres labelled with different symbols, some of them going back to the days of alchemy. This was of course extremely inconvenient and was shortly afterwards replaced by the letter and subscript system that we still use. However, the representation of atoms as sphere-like emphasises their physical reality, and if Dalton's system had survived it would probably have led to much earlier general discussion of how these different atoms are arranged in space.

A "compound body" is what we would call a compound. The "ultimate particle" of a compound body is what we would call a molecule. But what is the ultimate particle of a "simple body", an element? We have two different answers to this question; either a molecule, or an atom, depending on the exact context. Dalton made no such distinction, and assumed that the "simple bodies" of, for example, hydrogen or oxygen gas could not be further subdivided. And there the trouble started, and it took something like fifty years to sort out the resulting muddle.

Dalton went on to make what looks like a very reasonable assumption. If you have only one compound[3] between two different elements A and B, it will have the simplest possible formula, AB. This makes sense, given Dalton's explanation of gas pressure, which was based on the idea that particles of the same kind will repel each other. If a second compound exists, it will have the formula A_2B, or perhaps AB_2. Compounds of type ABC etc are also possible. There are further possibilities, such as A_3B and AB_3, but A_2B_2 does not appear in Dalton's catalogue.

The result was that Dalton assumed that water contains one ultimate particle of hydrogen and one of oxygen, ammonia one of nitrogen and one of hydrogen. Then, more or less in Dalton's own words, nitrous gas contains one nitrogen and one oxygen, nitrous oxide two nitrogens and one oxygen, "nitric acid" (meaning the gas formed when nitrous gas reacts with oxygen) one nitrogen and two oxygen, nitrous acid one nitric acid and one nitrous gas. One charcoal and one oxygen, carbonic oxide, but with two oxygen, carbonic acid. One charcoal and one hydrogen, olefiant gas. One charcoal and two hydrogen, carburetted hydrogen from stagnant water.

In our modern notation, water is OH (wrong), ammonia is NH (wrong), nitrous gas is NO (correct), nitrous oxide N_2O (correct), the gas formed between "nitrous gas" and oxygen, NO_2 (correct), nitrous acid N_2O_3 (this is indeed what you would get from nitrous acid if you took away the elements of water), carbon monoxide CO (correct), carbonic acid (again subtracting water) CO_2 (again correct), ethylene CH (should be C_2H_4), methane CH_2 (should be CH_4).

Dalton gets his ratios consistently right except when they involve hydrogen, where he makes one "ultimate particle" do the work of two or even three, and also gives (even after correcting this error) just half the right molecular structure for ethylene. After all, he had no reason to assume anything more complicated.

As we saw earlier (Chapter 2), this atomic theory immediately explains the rules of chemical combination. Any particular compound has a definite composition, and all the atoms of each element have their own definite weight, so this gives the *law of constant proportions*. When two elements form more than one compound, the proportions themselves are in simple proportion; this is the *law of multiple proportions*. But these laws tell us less than we really need to know about atoms, a point best illustrated by example.

Consider the two different compounds formed between hydrogen and oxygen. The most familiar, water, always contains hydrogen and oxygen in the ratio (by weight) of 1:8. Hydrogen peroxide (not actually discovered until 1818, but it shows the principle) contains the same two elements in the ratio 1:16. In other words, the ratios are themselves in a simple ratio.

This is very easy for us to understand. Everyone knows that water is H_2O. So one atom of oxygen weighs 16 times as much as an atom of hydrogen, and the atomic ratio of hydrogen and oxygen in hydrogen peroxide is 1:1 instead of 2:1, consistent with an "empirical" formula, based simply on the relative weights, of HO. In fact, the modern formula of hydrogen peroxide is H_2O_2, but this should not bother us, since all the relative weights can tell us is the ratio of the number of atoms, and 1:1 is the same ratio as 2:2.

The attentive reader may at this point have noticed a small problem. Water is the simplest compound between hydrogen and oxygen, and so Dalton gave it the formula HO. This would make each atom of oxygen just eight times as heavy as an atom of hydrogen, and hydrogen peroxide would have the formula HO_2. This fits the analytical proportions just as well, and Dalton's formula for water has the added virtue of simplicity. So how do we know that he was wrong about this?

Dalton himself speculated that more information might come from the densities of gases. Presumably, he reasoned, the density of a gas was determined by the mass of its ultimate particles. But he rejected this as a dead end. Steam, he knew, is less dense than oxygen at the same temperature and pressure, but surely the ultimate particle of water would have to be heavier than the ultimate particles of hydrogen and oxygen from which it was formed.

He also considered the case of "nitrous gas" (nitric oxide), much investigated at this time because of its ability to combine with oxygen. He correctly assumed that the formula for this gas is NO. He also knew that one volume of nitrogen combined with one volume of oxygen would give two volumes of NO. But if volume is proportional to the number of fundamental particles, this would imply that each fundamental particle of nitrous gas contained half a fundamental particle of oxygen, and half a fundamental particle of nitrogen. As long as the fundamental particle of an element is thought of as an atom, relating volumes to number of atoms leads straight to the need to consider half atoms as separately existing entities, which is absurd.

Notice, however, the curiously simple relationship between the volumes of gases involved. Just one volume of nitrogen reacts with just one volume of oxygen to give just two volumes of nitrous gas. Similarly, two volumes of hydrogen react with just one volume of oxygen and if we maintain the system above the boiling point, we end up with just two volumes of water vapour. As the French chemist Gay-Lussac demonstrated, the volumes of different gases involved in a chemical reaction are *always* in simple ratios to each other. For example, two volumes of what was then called "carbonic oxide" react with one volume of oxygen to make two volumes of "carbonic acid", substances that Dalton himself had correctly formulated as one carbon and one oxygen, and one carbon and two oxygens, respectively.

This is, to say the least, highly suggestive. According to Dalton, the ultimate particles of the elements combine in simple ratios. According to Gay-Lussac, gases combine in simple ratios. So perhaps there is some very simple relationship between the volume of a gas, and the number of ultimate particles that it contains. This much actually occurred to Dalton, but he rejected the idea as untenable, for the reasons spelt out in the preceding paragraphs. So Dalton dealt with Gay-Lussac's unwelcome result in exactly the same way that global warming denialists deal with the evidence of climate change; he refused to accept the accuracy of the experimental findings.

At this point, the atomic theory appeared to have reached an impasse. We can see this in the changing attitudes of one of Dalton's distinguished contemporaries, William Hyde Wollaston. Wollaston had a first-rate mind, wide interests, and a high degree of intellectual courage. An accomplished chemist, he devised ways of extracting platinum from its ores, and discovered two new elements (palladium and rhodium) in the process. He designed the Wollaston prism, manufactured from calcite, which lets through only polarised light. He was also the first to discover the dark lines in the Sun's emission spectrum (the so-called Fraunhofer lines), which give direct information about the chemical elements present in the Sun's corona; more on these in Chapters 5 and 15. As early as 1808, Wollaston had the idea of molecular *shapes*, determined by the arrangement of atoms within molecules, although he doubted if we would ever know what these actually were. He was interested in crystallography, and, in 1812, considered how spherical atoms could pack together and whether this could be related to the regular shapes of crystals. He even considered the best way of packing together two different kinds of atoms, using Dalton's rule that unlike atoms would attract, but like atoms would repel each other, and came up with the arrangement actually shown by the sodium and chloride ions in sodium chloride, common table salt.

By 1813, however, Wollaston had quietly retreated from reliance on Dalton's concept of atoms, adopting instead the more modest concept of "equivalents". For example, methane contains one part by weight of hydrogen to three parts of carbon. The gas then called carbonic acid contains eight parts by weight of oxygen to three of carbon. So one part of hydrogen is equivalent to eight parts of oxygen, and this in turn is consistent with the 1:8 ratio in water. As long as we only have weights to rely on, no more can be said, and a discussion of the ultimate and inaccessible particles from which matter is constructed remains beyond our reach.

Strange as this may seem to us, this unassuming approach held sway for something like 40 years. We have the curious spectacle of some of the best minds of their generation, including people like Wollaston and

Faraday, shying away from what seems to us the obvious, and not answering the question whose resolution seems to us now to form the very backbone of the subject.

I would propose what seems to be quite a shocking explanation. *It didn't really matter.* As long as there were so few elements that there was no real hope of observing patterns in their behaviour, and as long as there was no prospect of relating composition to what we would now call molecular geometry, nothing much really depended on whether the hydrogen in water corresponded to one atom, or two.

I spoke of the "backbone" of the subject. The backbone is absolutely integral to the body plan of vertebrates, but that is because of the super-structure (skull, ribs, etc.) that has developed around it. When the back-bone first appeared, it was in the barely recognisable form of a semi-disposable stiffening of the body plan, so that the maturing sea squirt can outgrow its rudimentary backbone as readily as Wollaston chose to outgrow the atomic hypothesis.

We owe the solution of the "half a fundamental particle" problem to one Lorenzo Romano Amedeo Carlo Avogadro di Quaregna e di Cerreto, Count of Quaregna and Cerreto in the Kingdom of Piedmont and Sardinia, sometimes known as the Kingdom of Savoy. He is for rather obvious reasons usually just referred to as Avogadro, as in Avogadro's Hypothesis (truly his own), and Avogadro's Number (nothing to do with him whatsoever).

Avogadro was born in Turin, capital of the Kingdom, in 1776. At that time, Piedmont was an absolute backwater. Italian unification was almost a century away, and Piedmont was as much French as Italian (indeed, the vocabulary of the Piedmontese dialect is still intermediate between these two). The king, Victor Amadeus III, who reigned from 1773 until his death in 1796, was an extreme reactionary paternalist, who reintroduced the Inquisition, and made Easter Communion and meatless Fridays com-pulsory for all his subjects. The country remained predominantly agricul-tural, with manufactured goods needing to be imported. In 1800, Piedmont was annexed by Napoleonic France, of which it remained part until 1814. With the French defeat, the victorious allies adopted a policy of putting the clock back to the pre-revolutionary era, and restored the House of Savoy to the throne.

Piedmont did however have a capable civil service, of which Avogadro was part during the 1790s, when he was what we would now call a legal aid lawyer. In 1804, Avogadro became a corresponding member of the Academy of Sciences of Turin, and, in 1820, he was appointed Professor of mathematical physics. In 1821, an insurrection in Turin demanding a constitution was bloodily suppressed with the help of Austrian troops. Over 70 people were executed, and Avogadro's chair was abolished as part of general reprisals against the University. However, the chair was reinstated 12 years later and he was eventually reappointed to it. He thought of himself as a physicist, but his hypothesis, put forward in *Journal de Physique* in 1811 and 1814, is now part of every introductory course in chemistry.

Ironically, this was virtually identical with the idea that Dalton himself had rejected many years earlier, that *different gases under the same conditions contain the same number of molecules per unit volume*. The only difference, and it is a crucial difference, is that Avogadro did not flinch from the "half-particle problem". Indeed, he dealt with it in exactly the same way that we do today. One molecule of nitrogen reacts with one molecule of oxygen to give two molecules of "nitrous gas". It follows that, to use Avogadro's terms, the "integral molecules" of the compound, and the "molecules" of the elements nitrogen and oxygen, each contain (at least) two "elementary molecules". Why only two, rather than four or six? Because, in all the reactions that have been carried out with these elements, there is never any need to suggest more.

The idea is beautiful and simple, but the language in which it is expressed makes it almost incomprehensible.[4] Let me say the same thing again, in modern terms:

The molecules of nitrogen and oxygen each contain (at least) two atoms. The molecule of nitrous gas contains just one atom of each, or, using our present system for writing chemical equations,

$$N_2 + O_2 \rightarrow 2NO$$

Dalton's "elementary particles" of hydrogen or oxygen were not atoms at all, but molecules, and dealing with half an elementary particle does not really present a problem, since the elementary particle consists of two separate atoms.

We can now make sense of the other examples discussed so far. Since two volumes of "carbonic oxide" react with one volume of oxygen to make two volumes of "carbonic acid", this confirms again that each molecule of oxygen must contain (at least) two atoms. So indeed does the reaction between carbon and a restricted supply of oxygen, to give "carbonic oxide" in the first place. Writing the modern formulae for these reactions, and applying Avogadro's hypothesis to Gay-Lussac's observations on volume, we have

$$2CO + O_2 \rightarrow 2CO_2$$

with the carbon monoxide having been formed by the reaction

$$2C \text{ (carbon, solid)} + O_2 \rightarrow 2CO$$

To derive the correct formula for water, we need to know how many atoms of hydrogen there are in a molecule of hydrogen. As it happens, one volume of hydrogen will react with one volume of chlorine to give two volumes of the strongly acidic gas, hydrogen chloride. It follows that both hydrogen and chlorine must have at least two atoms in each molecule.

Recall now that two volumes of hydrogen react with one volume of oxygen to make two volumes of steam. With the number of molecules being proportional to the volumes, we can write

$$2H_2 + O_2 \rightarrow 2H_2O$$

which is obviously (to us, with all the benefit of hindsight) the correct formula. Avogadro himself pointed out that "the hypothesis, supposing it well founded, puts us in a position to confirm or rectify his [Dalton's] results from precise data, and, above all, to assign the magnitude of compound molecules according to the volumes of the gaseous compounds, which depend partly on the division of molecules of which this physicist [Dalton] has no idea at all" and "the division of the molecule, which does not enter into Dalton's calculations, here again corrects in part the error that would result from his other suppositions." Once we recognise that hydrogen exists in the gas phase as H_2, we are led to conclude that the molecule of water contains two hydrogens, not just the single one required by Dalton's arbitrary assumption of maximum simplicity.

It would seem that Avogadro's hypothesis (as we still call it) was an idea whose time had come. More examples of reactions between gases were accumulating, and confirming that these reactions involved simple ratios of volumes. Dalton himself had thought of it but rejected it, and Ampère and Dumas in France put forward similar ideas independently. So why did it take another 50 years for the hypothesis to gain general acceptance?

We can identify five distinct reasons for this curious delay. The word "molecule" was, as we have seen, loaded with assumptions. The hypothesis required two identical "elementary molecules" (i.e. atoms) to bind together, although the pressure of gases was explained by saying that identical particles repel each other. This particular difficulty was actually made far worse by the work of Davy and Faraday, who showed how compounds such as common salt could be broken apart by passage of an electric current, and correctly inferred that it was the *difference* in a fundamental electrical property that was responsible for binding them together (more on this in Chapters 9 and 12). If chemical bonding depends on electrical differences, how is it possible for two atoms with identical electrical properties to bind together? This problem was only solved by the use of quantum mechanics, in the 1920s. Thirdly, Avogadro's own choice of words (which was not always consistent) was confusing and obscure. Fourthly, chemists were busy enough finding out the composition of things, and, as we have seen, "equivalent weights" were good enough for this purpose. And finally, worst of all, if Avogadro was right then the highly (and correctly) esteemed Dalton was wrong.

It is not so unusual for a completely incorrect theoretical assumption to outlive its usefulness because of the reputation of its proponent. I have seen two examples of this in my own career. When I first studied chemistry, bonding between what are called the "transition metals", metals like chromium, iron, and copper, and non-metals was described in terms of the same theory used (as in Chapter 12) for bonding between one non-metal and another.[5] This theory had the backing of Linus Pauling, one of the greatest chemists of his generation, and persisted until the late 1950s despite the fact that it gave completely inaccurate predictions and that an alternative theory[6] existed and had indeed been shown, decades earlier, to work. My second example is the theory, which persisted right up until the analysis of moon rocks, that the Earth had originally been formed in a cold state, and only slowly acquired a molten core as a result of the heat released through radioactivity. (This heat does indeed contribute in an extremely important way to the Earth's energy budget; see Chapter 1.) This view was not compatible with the obvious fact, clear almost a century earlier to Lord Kelvin, that enormous amounts of gravitational energy must have been released when the Earth came together, and that there was no way in which this energy could have rapidly been lost. However, the "cold accretion" theory was supported by Harold Urey, a giant among geochemists, and I think the rest of us just assumed that we weren't smart enough to follow his reasoning.

It did not help that Avogadro hardly travelled beyond the provincial backwater where he worked. It is an unfortunate accident that his two key publications appeared in a French journal at the height of the Napoleonic wars between France and Britain, since Britain was where the study of gases was most highly developed. A further complication was his attempt to generalise his results on gases, and apply them to solids. We now know that this was deeply misguided. Gases behave as they do because they consist of 99.9% empty space, so that the volume of the actual molecules (as opposed to their number) has very little effect on their properties. In the solid, however, the units are packed together (as Wollaston had surmised) pretty well as tightly as possible.

Avogadro himself made little mention of his hypothesis in later years, using instead the current language of chemical equivalents. However, he did recognise that chlorine was an element, not a compound, and made a reasonably accurate estimate of its molecular weight from its density. This was an important contribution, because it implied that hydrogen chloride was an acid although it contains no oxygen, something that Lavoisier had regarded as impossible. Avogadro retired from active teaching in 1850, but continued to publish on various minor matters until 1853. His obituary in 1856 mentioned his work on specific heats, atomic volumes, and electrochemistry, but made no mention of the hypothesis for which he is now remembered.

If you put together the ideas of Lavoisier, Dalton, and Avogadro, you come up with our own modern idea of the atom, and a good way to determine the relative atomic weights[7] of the different elements, provided some of their compounds are gases. For example, it was known (as explained above) that water contains hydrogen and oxygen in the ratio 1:8 by weight; using Avogadro's reasoning we conclude that the correct formula for water is H_2O, and putting these two facts together we see that one atom of oxygen weighs as much as 16 atoms of hydrogen. So oxygen has a weight of 16 on a scale (good enough for our present purposes) where we give hydrogen a weight of 1. But what about those elements, mainly metals, all of whose compounds were solid? For example, the metal manganese, first isolated in 1774, occurs in Nature as an oxide, which reacts with carbon to give carbon dioxide and the free metal. By comparing the weight of the oxide with the weight of the metal formed from it, it could be found that just under seven parts by weight of metal were combined with four parts by weight of oxygen. So what does this tell us? If we applied Dalton's principle of maximum simplicity, we would assume a formula MnO, and an atomic weight for manganese of around 28 (28 for Mn to 16 for O cancels down to the observed 7 to 4). But we know, from the example of water, that Dalton's principle cannot be relied on. What if the formula is Mn_2O, with an atomic weight of 14? Or, indeed, MnO_2, with an atomic weight close to 56 (the correct answer, as we now know)? If we had any independent method, even an approximate one, of estimating the atomic weight, we could then choose between these possibilities.

Sometimes, people do the right thing for the wrong reason. The poet T. S. Eliot did not have a high opinion of this way of proceeding,[8] but scientists have no such qualms. In 1818, the French physicists Pierre Louis Dulong and Alexis Thérèse Petit found, for some 13 elements where the atomic weight was known, that if you multiplied the atomic weight by the specific heat (the amount of heat needed to increase the temperature of 1 g of material by 1°C), you always got roughly the same answer. Like most of their immediate predecessors and contemporaries, including Lavoisier and Dalton, they considered that heat was a kind of material substance, going by the name of "caloric", and rationalised their result by saying that an equal amount of caloric was required to heat up an atom of any element. Heat, of course, is nothing of the kind; as discussed below, it is simply disorganised energy. However, the statistics of how energy can distribute itself among different atoms does indeed lead to the conclusion that (at high enough temperatures, at least) *equipartition* of energy will occur, meaning that different kinds of atoms will be associated with the same amount of heat. The rule itself is not exact,[9] and accurate measurements of heat capacity are not easy, but all we need is a rough answer that enables us to pick the right value from the range available.

Another ingenious way of establishing a relationship between different elements was developed by the German scientist (and expert on Persian language and history) Eilhard Mitscherlich. His idea was that there should be some relationship between the architecture of the individual molecules of a substance, and the shape of its crystals. So if two different elements gave rise to compounds with similar crystalline form, this might indicate a similarity between the way in which they form chemical combinations. Mitscherlich found such similarities of crystal shape in parallel with similarities of chemical composition, when studying phosphates and arsenates. To take an example that many people will have come across, it is very easy to grow beautiful octahedral colourless crystals of potash alum, potassium aluminium sulphate. Chrome alum, potassium chromium sulphate, gives deep purple crystals of identical shape, and indeed, it is quite easy to grow an overlay of chrome alum on a potash alum crystal. This is good reason to suspect that the two different kinds of alum resemble each other, and that aluminium can therefore show chemistry similar to that of chromium. The most common oxide of chromium can be converted to chromium metal by treatment with carbon, and shown (by determining the percentage of chromium by weight, and then using the formula of Dulong and Petit to choose between the implied possibilities) to have the formula Cr_2O_3. Aluminium oxide cannot be converted to metal in the same way, because the bonding between aluminium and oxygen is too strong, but you could argue by analogy for a formula Al_2O_3, and you would be right.

But back to Avogadro's hypothesis — and to what turn out to be two closely related topics — the nature of heat and the properties of gases. Newton had thought of heat as a kind of vibration, but Dalton, like many of his contemporaries, thought of it as a substance. (Remember that Lavoisier actually included "caloric" in his table of elements.)

This view of heat was challenged by Benjamin Thomson, Count Rumford of the Holy Roman Empire, a Loyalist (i.e. pro-British) refugee from the newly independent United States, who had placed his considerable talents at the disposal of Prince Karl-Theodore of Bavaria. Rumford found, while boring holes in metal to make cannons, that the metal got hot, and correctly inferred that heat was being generated by working on the metal. However, quantitative observations are much more convincing than the merely qualitative. James Prescott Joule, Dalton's most distinguished (indeed, only distinguished) student, was responsible more than anyone else for establishing that heat is a form of energy, by measuring what was called "the mechanical equivalent of heat". This he did by using falling weights to drive paddles inside a container of water, and measuring the increase in temperature. So heat was not a separate substance after all, but just a disorganised form of energy. William Thomson, Lord Kelvin, whom we met in Chapter 1, was very impressed by this result, which he incorporated into his formulation of the law of conservation of energy (First Law of thermodynamics), and became Joule's long-term collaborator. It is Kelvin who tells us that Joule took a large thermometer on his honeymoon in Switzerland, to measure the temperature of water above and below a waterfall. What Mrs Joule thought of this proceeding is not known.

So if heat is energy, how is that energy stored when stuff is heated? Regarding gases, Joule originally thought that it all went into making the molecules spin faster,[10] but in 1848 heard about and accepted the correct view, that at higher temperatures the molecules also move faster, increasing the number and intensity of their impacts against the walls of their container. Using this concept, he was able to calculate the speed of a hydrogen molecule (a little under 2 kilometers per second) with reasonable accuracy. This kinetic explanation of heat explains why pressure increases when you heat a gas in a sealed vessel; the molecules are moving faster, so each of them bounces off the wall more often, and in addition each bounce entails a larger transfer of momentum. It also explains exactly how it comes about that gases cool when they expand inside the cylinder of a heat engine. They are doing work against the piston, and bounce off it with reduced energy as a result. It had been thought that the cooling was due to the simple act of expansion, but Joule showed that this was not true by a beautifully simple experiment. He connected two vessels together through a stopcock, evacuated one of them, and pumped in air into the other one until the pressure was 20 times atmospheric. He then opened the stopcock, allowing the gas to expand to twice its volume, and showed that this led to no change in temperature. It is not expanding as such that cools down the working gas in a heat engine, but expanding against a load, and doing work in the process. The importance of this insight for the design of engines is obvious.

Using this *kinetic theory of gases*, it is quite easy to show that pressure is proportional to the average kinetic energy of the gas molecules.[11] But this average kinetic energy turns out to be directly proportional to temperature above *absolute zero*. It also turns out that at any given temperature, molecules of different gases have the same average kinetic energy. So different gases at the same temperature, with the same number of molecules per unit volume, will exert the same pressure. We can turn this last statement inside out. If two different gases at the same temperature are exerting the same pressure, then *they must have the same number of particles per unit volume*. This argument raises Avogadro's hypothesis to a whole new level. It is not just an *ad hoc* suggestion put forward to explain the laws of combining volumes of gases. It is an inevitable consequence of the statistical treatment of gases.

Meantime, the number of chemical examples illustrating Avogadro's principles was increasing, but Avogadro's own papers remained shrouded in obscurity. In 1857, Raffaele Piria argued for the diatomic nature of hydrogen on purely chemical grounds. It is formed by the reaction between copper hydride (discovered in 1844) and hydrochloric acid, and Piria reasoned that since the hydrogen produced contained material from two very different sources, each molecule of hydrogen had to contain two separate atoms. What is remarkable is that he did this work at the University of Turin, from which Avogadro had retired only seven years before, but did not seem to know that Avogadro had made the same suggestion for different reasons 40 years earlier.

By 1860, nomenclature in chemistry had reached such a state of confusion that a conference was called at Karlsruhe to establish precise definitions of "atom, molecule, equivalent, atomicity, alkalinity etc". The sound bite version of what happened next is that Stanislao Cannizzaro argued in favour of a system based on Avogadro's hypothesis, and carried the day. It was not really quite that simple.

Stanislao Cannizzaro spent some time as Piria's assistant, before returning to his native Palermo in 1847, just in time to join a revolution against the Bourbon kings. When this failed, he was forced to flee to Marseille, but in 1855 secured a position as Professor of Chemistry at Genoa. Lacking laboratory facilities there, he turned his attention to teaching, summarising his thoughts on the subject in his *Sketch of a Course of Chemical Philosophy*, or *Sunto di un corso di Filosofia chimica*,[12] which must be one of the most influential sets of lecture notes ever. In 1860, after Garibaldi's success in overthrowing the Bourbons, he returned to Palermo as Professor, and in 1871 moved to Rome, where as well as his university position he became a senator of newly united Italy, and continued to lecture until shortly before his death at age 83.

The *Sunto* is a masterpiece of clarity and tact (we should remember that it was written in a period of fierce nationalisms, extending even to disputes in science). Ampère and Dumas are generously cited as independent originators of the hypothesis, the experimental results of numerous other chemists from France and Germany enlisted in its support, and the precise differences between their viewpoints and Cannizzaro's own carefully explained. From its very first sentence, the *Sunto* maintains a lucid distinction between molecules and atoms, stating that "the molecules of the different substances, or those of the same substance in its different states, may contain a different number of atoms, whether of the same or of diverse nature." There follows a detailed discussion of the laws of combining ratios of weights, how these follow from the concept of an atomic weight, and how the hypothesis links these facts directly to the combining ratios of volumes, provided the distinction between "atom" and "molecule" is maintained, both for compounds and for elements. However, he makes no appeal to the kinetic theory of gases, (although Clausius, one of its developers, is cited in support). This seems strange from our perspective, and reflects the poor level of communication between physicists and chemists at that time. Where gaseous state density data are unavailable (as in the case of carbon, and most metals and their salts, for example), Cannizzaro uses the observed fact that specific heat per atom is roughly constant, a generalisation of the rule of Dulong and Petit mentioned above.

One further merit of the *Sunto* is its choice of unit of atomic weight; half a molecule of hydrogen. Cannizzaro carefully avoids saying "an atom of hydrogen", to allow for the conceptual possibility that this half-molecule can be further subdivided, but to us his meaning is sufficiently clear. This is the lightest atom, and if we use it as our standard a remarkably large number of atomic weights turn out to be whole numbers, at least within the accuracy of measurement then available. (For more on this, see Chapter 5, The Discovery of the Noble Gases.) The present standard, which fixes the weight of one atom of C-12 at 12 units, is very close to Cannizzaro's, but is preferable for practical reasons of measurement (very many elements form compounds with carbon, which can be examined in the mass spectrometer), and, of course, avoids the vagueness that arises from the existence of isotopes, something of which Cannizzaro could have had no knowledge.

The Karlsruhe conference did not reach any clear conclusion, but gave Cannizzaro the opportunity to distribute copies of his *Sunto*, thereby convincing enough key players to ensure general acceptance. However, we know that as late as 1872, Cannizzaro had to labour the point in his lectures, showing that the controversy was still alive, and some chemists insisted on writing the formula of water as OH as late as 1890. Even later, some philosophers of science insisted that postulating physical reality for atoms and molecules went beyond the evidence, and the matter was only finally laid to rest in 1905, when Einstein explained Brownian motion (the random jiggling of very small particles suspended in water) in terms of the statistical fluctuations of molecular motion.

What about Avogadro's number? Cannizzaro held out no prospect of determining the actual weights of individual atoms, but all that mattered for his purpose was the ratio between them. Now, using the mass spectrometer, we can measure such things with great accuracy. Avogadro's number is defined as the ratio of the atomic mass unit to the gram, or, what amounts to the same thing, the number of atoms of carbon-12 in 12 g carbon-12. Its actual value according to our best current estimates is around $6.02214179 \times 10^{23}$, or 602,214,179,000,000,000,000,000 atomic mass units per gram. Since the atomic mass unit is also known as the *Dalton*, and is a refinement of the unit introduced by Cannizzaro, it would have made sense to call this number "Dalton's number" or even "Cannizzaro's number". Avogadro had nothing to do with it.

A Note On Homoeopathy

The German physician Samuel Hahnemann, an impressively long-lived contemporary of Lavosier, Dalton, and Avogadro, claimed in 1796 that diseases could be cured by natural remedies that mimic the effects of the disease itself. Thus cinchona bark, which causes sweating, is effective against malaria. Hahnemann also concluded that the greater the dilution, the greater the efficacy, because "succussion," the process of striking against a leather surface after each dilution, increases the effect of the treatment by a process of "potentisation". The most fully potentised, and therefore the most effective, homoeopathic medicines are used at dilutions of one in 10^{30}, at which point there will be (as we can see by comparing with Avogadro's number) on average one molecule in every sixteen thousand *tons* of the preparations, which are sold by the ounce.

At the time of writing, homoeopathy is still available on the National Health Service in Britain, although it is not certain whether it will survive the present round of budget cuts. However, there is no doubt that homoeopathic remedies will continue to be available from Boots, the leading pharmacist in the United Kingdom, and homoeopathy is endorsed by the heir to the throne, Prince Charles, who has been much criticised for using his influence on its behalf.

To Sum Up

The idea of separate particles of matter, atoms in the modern sense of the word, arose in a rather roundabout way from consideration of the solubility of gases. This led John Dalton to propose that different gases were made up of different "ultimate particles" (what we would now call molecules) that did not compete with each other for sites within the solvent. In much the same way, they did not, as gases, compete for space, so that each gas exerted its pressure independently of the others. Different kinds of fundamental particle corresponded to different compounds, or to different elements. The fundamental particles of elements were assumed to be indivisible atoms, and the composition of the ultimate particles of compounds was determined by the different numbers of fundamental particles of elements that they contained. This is the atomic theory of chemistry as we would now recognise it, although it contained a built-in confusion, as far as the elements themselves are concerned, between the concepts of "atom" and "molecule". This theory explained the laws of constant proportions and multiple proportions. In addition, Dalton assumed that the numbers of different kinds of atom in a molecule would be as simple as possible. Water, for example, would have the formula HO.

Gay-Lussac discovered that the volumes of gases taking part in, or produced by, a chemical reaction were in simple ratios. This led various people, including Dalton himself, to speculate that there was a fundamental reason for this, connected with the number of molecules in a given volume of gas. But Dalton rejected this explanation, on the grounds that it would require the ultimate particles of an element to be sub-divisible, which seemed to him absurd. Avogadro, working in obscurity in what is now northern Italy, realised that Dalton had failed to distinguish, in the case of elements, between atoms and molecules. He recognised that Gay-Lussac's results could be explained if molecules of the common gases hydrogen, nitrogen, and oxygen contained, not one, but two atoms, and if equal volumes of different gases at the same conditions of temperature and pressure contained equal numbers of molecules. This last suggestion is still known as *Avogadro's hypothesis*.

In the mid-19th century, physicists developed the kinetic theory of gases, according to which heat is simply disorganised kinetic energy, and deduced the truth of Avogadro's hypothesis as a consequence. Nonetheless, chemists as a group resisted this conclusion until persuaded by Cannizzaro's demonstration of its utility in generating a consistent set of atomic and molecular weights. Philosophically inspired scepticism about the reality of atoms and molecules persisted into the 20th century, and was only dispelled by Einstein's demonstration that the molecular hypothesis explained Brownian motion, the random jiggling of small particles suspended in solution.

5 The Discovery of the Noble Gases — What's so New About Neon?

"The most exciting phrase to hear in science, the one that heralds new discoveries, is not Eureka! (I found it!) but rather, 'hmm ... that's funny ...'"

Isaac Asimov

Sun-stuff, new, lazy, hidden, strange, gas-from-radium; helium, neon, argon, krypton, xenon, radon. These noble (meaning unreactive) gases are not on the whole particularly rare substances. Helium, in fact, is the second most common element in the Universe, after hydrogen, while the air in an average sitting room contains about a kilogram of argon. The first real evidence for these elements dates back to the eighteenth century, when the very concept of a chemical element was in its infancy, but it took a hundred years for this evidence to be properly appreciated. The noble gases seemed at first to violate all the known rules of chemistry, but came to play a central role in our understanding of the chemical forces that bind all substances together. The really interesting fact about them, scientifically speaking, is their almost total lack of chemical properties.

Hidden in Plain Sight

In 1783, the English scientist[1] Henry Cavendish sparked together mixtures of air and the gas we now know as oxygen, making products (nitrogen oxides) which dissolve in water. By adjusting the ratios of air and oxygen, he could get almost all the gases to disappear in this way. Almost, but not quite. In his own words, "... if there is any part of ... our atmosphere which differs from the rest ..., we may safely conclude that it is not more than 1/120 part of the whole." Cavendish presumably regarded this "1/120 part" as a minor matter; if so, he was doing less than justice to the significance of his own observations.[2]

An Extraterrestrial Element

In 1825, the French philosopher Auguste Comte said that there are some things that we will never know, among them being the chemical composition of the stars. This was an unfortunate example. We do not need actual material from the stars in order to analyse them. What we do need, of course, is information, and this they send to us plentifully, in the form of light. To reveal this information, we must separate the light into its different wavelengths or colours.

Many years earlier, Isaac Newton had passed sunlight through a glass prism, and found that this gave him all the colours of the rainbow. In 1802, William Wollaston in England, whom we met in the last Chapter, repeated this experiment, using an up-to-date high quality glass prism, and discovered that the continuous spectrum of colours was interrupted by narrow dark lines. (Rather unfairly, these are now known as the *Fraunhofer lines*, after German physicist Joseph von Fraunhofer, who confirmed and extended Wollaston's observations.) These lines later proved to match exactly the light given out by different elements when heated or in electric discharges. A now familiar example is provided by the element sodium, whose yellow emission is used in street lamps.[3] The lines can be used as a kind of fingerprint for each element. We can take this fingerprint using electric discharges here on Earth, and compare it with the lines found in the spectrum of the Sun. In this way, we can get a good chemical analysis of the surface layer of the Sun or of any other star whose light we can collect, and astronomers now extend the same process to the most distant galaxies.

The yellow emission of sodium in fact consists of two lines very close together. In 1868, Norman Lockyer was looking at these lines in the spectrum of the Sun's atmosphere when he saw a third line in the same region which was not part of any known pattern. It was by then well-established that every element gives rise to its own characteristic line spectrum, although the reasons for this were still completely mysterious. (We now know the lines to be caused by electrons jumping between different fixed energy levels, as described in Chapter 11, and therefore giving rise to photons of definite energy and thus definite frequency, but these concepts still lay far in the future.) Lockyer correctly inferred that a new line must be due to a new element, not yet discovered on Earth, and named this new element *helium* (Greek *helios*, Sun). Lockyer probably imagined that helium was a metal (hence the choice of a name ending in –ium, like sodium, magnesium, etc.) and that it had not been found on Earth because it is rare. Helium is in fact the second most common element in the Universe. According to current thinking, it was formed, together with hydrogen, in the few seconds after the Big Bang. It is also the first and principal material into which hydrogen is converted by the nuclear fusion reactions that power the stars, while on Earth it is formed by radioactive decay of the heaviest elements. Most of it escapes, but some gets trapped in the same kind of porous rock formation as natural gas.

"That's Funny ..."

The next major development came from a completely unforeseen direction. They usually do, which is part of what makes them major. It is also part of what makes science so difficult to manage, and so subversive of all complacent certainties, including its own.

John William Strutt, Third Baron Rayleigh, was among the last in the tradition of independently wealthy Englishmen pursuing science as a vocation, rather than a profession (another such was Charles Darwin). Senior wrangler (top of class in mathematics) at Cambridge, he was elected a fellow of Trinity College in 1866, but on the death of his father in 1873 he took up residence at the family seat in Essex, where he built a laboratory. From 1879 to 1884, he served as Cavendish Professor[4] of Experimental Physics at Cambridge, and from 1887 to 1905 as Professor of Natural Philosophy at the Royal Institution. Throughout his career he published some 450 scientific papers on a wide range of fundamental topics in every branch of physics. He found the answer to that most difficult of children's questions, why is the sky blue. He showed that this was because of what is still called *Rayleigh scattering*; shorter wavelength light (towards the blue end of the spectrum) is scattered more by fluctuating local concentrations of molecules in the atmosphere. He was also the father of the Strutt whom we met in Chapter 1, as Arthur Holmes's research advisor.

We saw in the last chapter how the combined insights of Dalton and Avogadro led to our modern concept of atomic and molecular weights,[5] and the realisation that the densities of different gases were directly proportional to the weights of the individual molecules of each gas.[6] In view of this, Lord Rayleigh was very interested in the suggestion known as "Prout's hypothesis". Prout, an Edinburgh-trained physician and part of the "Scottish Enlightenment" circle (see Chapter 1), had measured the densities of several different gases and noticed that his answer was often close to a whole number times that of hydrogen. This is indeed no accident. For any isotope, the atomic weight relative to hydrogen is very close indeed to the mass number (sum of number of protons and neutrons in the nucleus), since protons and neutrons weigh roughly the same as a hydrogen atom, and many elements, including hydrogen, carbon, nitrogen, oxygen, and fluorine, consist predominantly of one single isotope. Prout, knowing nothing of such sub-atomic particles, boldly speculated that hydrogen itself might be a basic building block from which the other elements were constructed. The densities of nitrogen and oxygen were close to what Prout's hypothesis demanded, but not spot on, and Rayleigh decided to re-measure these as accurately as he could.

There were (and are) two kinds of ways of obtaining nitrogen − by chemical reaction, and by isolation from the air. Rayleigh made his "chemical" nitrogen by decomposing ammonia (NH_3) into nitrogen and hydrogen over hot copper. It is then a simple matter to convert the hydrogen to water by reaction with copper oxide, and remove the water and any residual ammonia by passing the gas through concentrated sulphuric acid. The decomposition of ammonia had in fact already been studied by William Ramsay, then Professor at Bristol, who noticed one tantalising fact about it; the nitrogen–hydrogen mixture formed in this way always contained residual amounts of ammonia (more on this in Chapter 6). Rayleigh prepared "atmospheric nitrogen" by taking air and removing the oxygen and other known components. He then measured the densities of both kinds of nitrogen with high accuracy, confidently expecting to get the same value for both. He did not; "chemical" nitrogen was slightly, but measurably, less dense, by about one part in a thousand. This was completely unexpected, since there is no way in which the properties of an element could depend on how it was prepared. The difference persisted even after Rayleigh had carefully eliminated all obvious sources of error, for example by showing that spiking the chemical nitrogen with extra hydrogen, before the copper oxide treatment step that removed it, made no difference to the final result.

An Answer, but More Questions

At this stage, Rayleigh was so puzzled that he published a letter in the scientific journal *Nature* asking for suggestions. It so happened that Ramsay, by then Professor of Chemistry at University College London, had read an old biography of Cavendish while still a student, and remembered his "1/120 part" of air that had failed to react. He discussed the matter with Rayleigh, and with his approval studied the reaction of air with magnesium. Magnesium is a metal with low density and an excellent strength to weight ratio. When cold, it is reasonably unreactive, and is even used for specialist applications such as sporting bikes. When heated, it burns very brightly in air, giving out a brilliant flare that was at one time used for photographic flashes. It is one of the very few substances that reacts with nitrogen gas, as well as with oxygen, and if air were essentially nothing but a mixture of oxygen and nitrogen, reaction with hot magnesium would get rid of it entirely. Of course, this is not what happens. Like Cavendish, Ramsay found an unreactive fraction, and Rayleigh and Ramsay together made their findings public in 1894. They showed that the unreactive fraction of the air was a new gas, which did not react with sodium metal, sodium hydroxide, chlorine, oxygen or sulphur, and gave it the name *argon* (Greek: lazy) to denote this fact. They measured its molecular weight,[7] and found it to be 40 times that of a hydrogen atom, just what would be needed to explain the higher density of "atmospheric nitrogen".

You might think that this work would be immediately accepted and applauded. Not so. It was criticised on several different contradictory grounds. Maybe their material was just a new form of nitrogen, containing three atoms in each molecule instead of two, just as ozone is a form of oxygen containing three atoms per molecule instead of the usual two. This would very nicely explain why its presence makes the nitrogen denser. However, it would not explain why it occurred in air but not in "chemical" nitrogen, or why it was so unreactive.

Maybe the effect was due to a contaminant. James Dewar, inventor of the Dewar vacuum flask, thought so, and remarked (in letters to *The Times*)[8] that he had often noticed a white material when carrying out his pioneering work on liquefying air. Dewar was perfectly correct in his observations, but completely wrong in his interpretation of them. With hindsight, it is easy to see that his "contaminant" was in fact solid argon, and that it offered confirmation of the new material, rather than the reverse. By his ungenerous response to Rayleigh's and Ramsay's discovery, Dewar merely cheated himself out of the credit for being the first to prepare argon as a solid.

There was, however, a far weightier objection. Chemists had now been cataloguing the elements for a century, and believed, with good reason, that with a few gaps among the rarer metals the catalogue was more or less complete. If argon really was a new element, with new properties, present in quite appreciable quantities, this implied not just one but a whole family of gaps in the catalogue.

Throughout the late 19th century, chemists had been busily classifying the elements according to their properties and relative weights. Mendeleev in Russia, among others, developed the *Periodic Table*, which arranges the elements in rows ("periods") that more or less correspond to increasing atomic weight, in such a way that elements with similar chemical properties end up in the same column, or "group". If he knew about the discovery of helium in the Sun, he was probably not at all worried by it. His Table contained plenty of blank spaces corresponding to elements that had not yet been discovered, and he would no doubt have just assumed that helium was one of these. So far from being worried by the prospect of new elements, Mendeleev was busy predicting with reasonable accuracy their properties and relative weights. In particular, he predicted the existence and properties of the elements needed to fill the spaces in his table underneath aluminium and silicon, and was proven right within a few years when these elements (which we now call gallium and germanium respectively) were isolated. It was this ability to predict that led to the triumphant acceptance of his Table. However, there seemed to be no place for a new element with the mass and properties of argon, and Mendeleev was among those who initially argued in favour of the "nitrogen ozone" formulation.[9]

The evidence from the new element's line spectrum, however, was quite conclusive. Line spectra can be observed for gases when a voltage is applied across a tube full of the gas at low pressure, and the line spectrum of any material is an *atomic* property (you will remember the importance of line spectra in the discovery of helium). For example, ozone has the same line spectrum as ordinary oxygen, even though it contains three atoms per molecule instead of the normal two. So if the new gas really were a new form of nitrogen, it would have the same line spectrum as the familiar form. Clearly, it did not. Its line spectrum was completely new; therefore, it could only be a new element.

Additional, equally conclusive evidence came from the specific heat of the new gas, the amount of heat energy required to increase its temperature. A gas that contains two or more atoms in each molecule can store energy in three ways; as the kinetic energy of motion of the molecule as a whole, as rotational energy (think of the molecule as a spinning dumb-bell), and as vibrational energy (think of this dumb-bell as held together by a vibrating spring). The new element had an unusually low specific heat, just the specific heat, in fact, that corresponded to the kinetic energy alone. Only one other substance was known at the time that had so low a specific heat, and that was mercury vapour, which was known from its density to have only one atom per molecule. Unlike hydrogen (H_2), oxygen (O_2), nitrogen (N_2), chlorine (Cl_2) and so on, this element was so unreactive that it would not even make chemical bonds with itself.

Six for the Price of One

This was extremely serious. If it was a new element, how could it be fitted into the Periodic Table? Since it had an atomic weight of 40, it would have to fit in somewhere near calcium (atomic weight 40.08), and, if the Periodic Table was to survive this intrusion, the new element would have to be part of a completely new, and hitherto unsuspected, group. But there was no known group of elements in the right position, or resembling argon in its total absence of chemical properties. This meant that Rayleigh and Ramsay must have discovered, not just one element, but the first member of a whole new family of highly unreactive elements, the *noble gases*. At first, and for many years afterwards, they were called the *inert gases*, in the belief that they formed no compounds at all. However, it would eventually be discovered that xenon and to a lesser extent krypton and even argon (just about) do form a handful of compounds with the very reactive elements fluorine, oxygen, and chlorine.

If the new group of elements was real, it should be possible to find and isolate the other members, and over the following five years, Ramsay did just this by careful fractional distillation of the unreactive portion of air. Argon was by a long way the main component; we now know that this is because it is formed by the decay of potassium-40, a naturally occurring isotope of potassium that is present as one part in 10,000 in potassium-containing rocks. The lightest member could not be isolated from the atmosphere, but is found trapped in radioactive minerals. Its spectrum showed it to be the same as helium, already named after the Sun. Some years later, Rutherford was to prove that its nucleus was the same as the alpha particles emitted during radioactive decay, thus solving the mystery of how this unreactive gas found its way into the rocks in the first place.

Over the next few years, Ramsay isolated neon (the new one), krypton (the hidden one), and xenon (the stranger) by fractional distillation of air, something now carried out on a massive scale. The main products, nitrogen and oxygen, are used in a range of industrial processes, while argon is used, because of its inertness, in gas-filled light bulbs, and neon, because of its strong red emission lines, in electrical displays.

Helium generated in rocks tends to get trapped in the same kind of geological formations that trap natural gas. In fact, 80% of the world's helium production comes from the oil and gas fields around Amarillo, Texas. Helium remains rare in the atmosphere, however, because its low mass allows it to escape from Earth gravity. It is used in balloons and dirigibles, as the refrigerant for the superconducting magnets used in MRI, and, because of its low solubility in blood, in breathing gas for divers.

The last member of the group, radon, was discovered by Ernest Rutherford in 1899 when he found that part of the radioactivity of an ore could be swept away in a stream of gas. It is formed by the radioactive decay of radium, and is the same radon gas that causes health concerns, especially in houses built over granite and other uranium-containing rocks. It is itself radioactive, but the main danger comes from its immediate decay products, solids which can attach themselves to household dust and end up lodged in the lungs. The biggest risk is of lung cancer, and the effect is greatest in groups whose lungs have already been exposed to mutagens, such as smokers.

Are there any more new elements waiting to be discovered? Yes and no. We know that each element corresponds to a whole number (equal to the number of protons in its nucleus). The task of identifying all the elements up to 92 (uranium, the highest atomic number to occur naturally on Earth) has long since been completed, but we have known how to make the even heavier elements, neptunium and plutonium, since the days of the Manhattan Project. The search for yet heavier elements than these involves bombarding heavy elements with complete nuclei, using powerful accelerators such as the one at Dubna, in Russia. The highest atomic number element so far produced is Ununoctium, the unlovely temporary name for Element 118. This is also predicted to be, chemically speaking, a member of the noble gas group, but it is so short lived and so difficult to produce that there is no prospect of examining its chemical properties.

Implications

To chemists, the most significant feature of the noble gases is their lack of chemistry. The noble gases have a secure home in the modern Periodic Table, in between the halogen elements fluorine, chlorine, bromine, iodine, and the alkali metals sodium, potassium etc. Terrestrial argon itself is somewhat anomalous, being one of a handful of elements whose atomic weight is greater than that of the element immediately following it in the Table. This is a result of its formation from potassium-40, which is heavier than the more common stable isotope, potassium-39. In the Solar System as a whole, the predominant isotope is argon-36, which would sit comfortably between chlorine, main isotope chlorine-35, and potassium-39.

Ironically, the problem that first caught Rayleigh's attention turns out to be of no great fundamental significance. Ordinary nitrogen is a mixture of two different isotopes, kinds of atom that differ in mass but have the same number of protons and electrons and, therefore, almost identical chemical properties. The measured atomic weight of nitrogen depends on how much of each isotope is present in the mixture, and this, in turn, depends on the precise details of the processes that led to the formation of Earth. The ratio of isotopes can and does differ elsewhere in the

Universe; in fact, nitrogen on Mars is more enriched in the heavier isotope, which is one of the ways that we can recognise the meteorites that originated there.

The Periodic Table, on the other hand, is central to our understanding of the chemical properties of all matter. Perhaps, in some far part of our galaxy, there is another hospitable planet, with sunlight, air and water, on which intelligent life has evolved to the point of making scientific discoveries. If so, I cannot even guess at the appearance and biochemistry of its chemists, but I am confident that they will have, hanging on their classroom walls, written in their own symbols, the same periodic table that hangs on ours.

One final implication, more important than all the rest. Much of the behaviour of the other elements can be understood as a tendency, by losing, gaining, or sharing electrons, to achieve the stable electronic arrangements found in the noble gases. From the early 19th century onwards, it had become increasingly clear that bonding in salts was electrical in nature; that potassium chloride, for instance, was an ionic compound, giving separate K^+ and Cl^- ions in the melt or in solution. This explained why such compounds were electrically conducting in solution or when molten, and why the passage of electricity led to their decomposition.[10] The electron itself was discovered in 1896 (see Chapter 2), and the implication was that K^+, Ar, and Cl^- were *isoelectronic*. If the electron distribution around the noble gases is particularly stable, then this accounts for their low reactivity, and for the charges of the ions formed by the groups on either side of them. So the positions of the noble gases in the Periodic Table, and the intervals between them, are no accident. As we now know, these intervals correspond to the numbers of different ways in which electrons can be packed round the central nucleus of an atom, giving successive stable layers with 2, 8, 18 or 32 members. These numbers, in turn, correspond to different sets of solutions to the equations of quantum mechanics, the theory discussed in Chapters 11 and 12 that describes the behaviour of very small particles, such as electrons. By the early 1900s, the American chemist G.N. Lewis (of whom we will hear much more in Chapter 12) had taken the next step, with his idea of electron sharing in covalent bonding, and of the stable octet. This is an essentially correct account of the nature of the chemical bond, all the more remarkable for having been first thought of a quarter century before quantum mechanics, and the theory of the electronic wave function, which shows how this sharing occurs.

To Sum Up

Around 1% of the air we breathe is actually a radioactive decay product, argon-40, formed by the decay of potassium-40 over geological time (and further proof, if such were needed, of the antiquity of our planet). This escaped notice for a century because of its unreactivity, until it was (re) discovered as a result of its effect on the density of "atmospheric nitrogen". Originally a major conundrum to chemical theory, it turned out to be one of a whole family (a "group") of highly unreactive gases, whose reluctance to form chemical bonds provides crucial insights into the very nature of chemical bonding.

6

Science, War, and Morality; The Tragedy of Fritz Haber

Tragedy is a word that is much misused. It should be reserved for disaster that overtakes a hero as a result of his greatness, as well as his failings. Hamlet is a tragedy, not just because our hero ends up dead, taking half a dozen people with him, but because his own reflective intelligence is instrumental in his fate. By this strict definition, the story of Fritz Haber, indicted war criminal, Nobel laureate, patriot and miserable exile, is indeed a tragedy. He sought to serve his country, and helped destroy it. The moral dilemmas of Haber's career will not go away, and the ironies of unintended consequences are timeless.

Early Life and Career

Fritz Haber[1] was born in 1868, in Breslau, then part of Prussia and, after Germany was unified in 1871, one of its major cities. But you will not find Breslau on the map today. In its place, you will find Wrocław, a Polish city whose German inhabitants chose or were persuaded to leave in the years after 1945, and Haber's own actions, as we shall see, were indirectly in part responsible for this transformation.

His father, Siegfried Haber, was a successful dyestuffs merchant, specialising in the importation of pigments, dyestuffs, and natural indigo. His mother died giving birth to him, but when Fritz was 9, his father remarried to a much younger woman, of whom Fritz was very fond. Haber's relations with his father were always difficult, and in his earliest years he spent much time with other relatives, but he was very close to his half-sisters, one of whom (Else) was to be his supportive companion, after she was widowed, towards the end of his life.

Fritz spent time at the Universities of Berlin, and Heidelberg, where he studied under Bunsen, of Bunsen burner fame. In Heidelberg, along with the customary dueling scar, he acquired a clear recognition of the importance of experiment in chemistry, and of its underpinnings in physics and mathematics. This was followed by a year's compulsory military training, which was not perhaps too onerous. He was garrisoned near his home, and had time to attend lectures in philosophy at the University of Breslau.

In 1889, he started graduate work at the Charlottenburg *Technische Hochschule* (Technical University). This was a major engineering school in Berlin, with 150 Faculty and 4,000 students. He worked with the organic chemist Carl Liebermann on the hot new topic of synthetic dyestuffs, followed by a Ph.D. dissertation on piperonal, one of the flavouring agents produced by the vanilla plant. In 1891, however, he achieved a mediocre grade in his Ph.D. oral examination, creating a poor impression by his inability to answer an elementary question about electrolytes (salts dissolved in water).

In Germany in 1891, there were more Ph.D.s graduating than there were jobs for them. Fritz took temporary jobs related in various ways to chemistry, and then went to work for his father. This was the time when synthetic dyes were in the process of replacing natural dyestuffs (remember Fritz's first research topic), and Siegfried probably hoped that Fritz would succeed him in the family business. However, in 1892, when cholera broke out in the port of Hamburg, Fritz persuaded his father to invest heavily in calcium hypochlorite disinfectant. But the purchase was mistimed, the business lost money, and Siegfried finally agreed that Fritz should make his career in academics.

After more short-lived positions, Fritz found a post at Karlsruhe *Technische Hochschule,* as an assistant in the Department of Chemical and Fuel Technology, which was at that time at the forefront of work in this area. His work there on the refining and combustion of hydrocarbons was published as a book in 1896, and earned him his *Habilitation*, qualifying him for a formal academic position. This is not quite as good as it sounds;

the position was that of *Privatdozent*, an instructor who was paid according to the number of students he could attract.

At Karlsruhe, he took up the study of electrochemistry, as if to undo the humiliation of his Ph.D. oral, and showed that the course of the electrochemical reactions of nitrobenzene depended on the applied voltage. (A good problem; details of this reaction were still being studied a century later). This led to another book, *Outline of Technical Electrochemistry based on Theoretical Foundations* (note the combination of practical and fundamental), which earned him tenure and promotion to Associate Professor, a more dignified title in Europe than it is in the US. Promotion to Full Professor came in 1905, after the publication of *Thermodynamics of Technical Gas Reactions*; note again in this title the connection between practical application and underlying fundamental science.

In 1901, Haber, now in a position to marry, resumed his courtship of Clara Immerwahr, whom he had first met 10 years earlier, and they married in August of that year. Clara had been the first woman ever to obtain a Ph.D. in Chemistry at the University of Breslau. Their marriage was initially happy, but Clara became increasingly depressed with a domestic role at once demanding and confining, while Fritz would return from work tense and impatient. Later, when Fritz was at the height of his career, Clara was to write to her friend and former teacher Richard Abegg, "What Fritz has gained in these eight years, that — and even more — I have lost."

Like Fritz, Clara had been born Jewish; like him, she had converted to Christianity. This was not unusual among German Jews at the time; the Jewish-born poet Heinrich Heine referred to baptism as "a passport into European culture." Haber's own physician came from a family that had converted, and Haber served as godfather at his son's baptism. Religious doctrines as such were probably not taken too seriously, and the idea was that in the enlightenment of the approaching 20th century, such things didn't really matter. Besides, there was much discussion at the time of what would be involved in forming Germany, only recently unified, into a nation. In this context, it is not surprising that the esteemed historian Theodore Mommsen (Nobel Prize for literature, 1903) wrote — while defending (!) Germany's Jews from anti-Semitic attacks — that they would have to make appropriate cultural adjustments or "face the consequences".[2]

Among Haber's talents was a gift for finding and attracting outstanding colleagues. His early electrochemical studies were carried out in collaboration with his friend Hans Luggin, inventor of the Luggin capillary still used in electrochemical measurements. This collaboration ended with Luggin's early death in 1899. Later, Haber worked with Zygmunt Klemensiewicz on the development of the glass pH electrode, still used routinely for measuring the acidity of solutions. (In addition to his reputation as a chemist, Klemensiewicz is still remembered as a mountaineer, pioneering the sport in the Tatra Mountains of his native Poland.) Friedrich Bergius, another colleague of the Karlsruhe years, studied the reactions of gases at high pressure, work that would eventually lead to the first process for the conversion of coal to oil, a reaction of great strategic importance, as we shall see, and to the sharing of a Nobel prize with Carl Bosch, whom we meet later in this story.

Haber always argued that one important function of knowledge was to be applied, but that at the same time technology needed a sound scientific basis. To quote his own Nobel Prize lecture, he believed that "while the immediate object of science lies in its own development, its ultimate aim must be bound up in the moulding influence which it exerts at the right time upon life in general and the whole human arrangement of things around us." Karlsruhe was (and is) closely associated with local industry, and in particular with the chemical giant BASF, *Badische Anilin und Soda Fabrik*, at that time the largest chemical company in the world. Throughout his career, Haber supplemented his income by industrial consulting, and his work with BASF was to lead to the achievement for which he will always be remembered.

Simple Reaction, Complex Problem, Major Importance

At this stage, as an expert on thermodynamics, and on the industrial reactions of gases, Haber was naturally drawn to the outstanding chemical problem of the decade: the formation of ammonia directly from its elements nitrogen and hydrogen. This means taking one molecule of nitrogen (N_2) and three of hydrogen (H_2), and reacting them together to make two molecules of ammonia (NH_3). This ammonia can be converted to nitric acid and nitrates, using processes already well established at that time.

The principal use of nitrate was as a fertiliser. This had been known for centuries, and nitrate use was fundamental to the agricultural revolution of the mid-19th century. In 1900 the main source of nitrates and nitric acid was Chile saltpetre, naturally occurring impure sodium nitrate derived ultimately from bird and bat excrement. This material, which had formed vast accumulations under the extremely dry conditions of the Atacama Desert, provided 60% of all nitrogenous fertiliser worldwide, and Germany was importing one third of all Chilean production.

The Chilean deposits are extensive, but not unlimited, and their essential role in agriculture was already foreseen as a serious long-term problem. In 1898, William Crookes, President of the British Association for the Advancement of Science, had warned that some way must be found to convert atmospheric nitrogen to nitrate, saying "It is the chemist who must come to the rescue... It is through the laboratory that starvation may ultimately be turned to plenty." But that was not Haber's main motivation.

Nitric acid is an essential raw material for the production of explosives, which rely on the rapid reaction between combustible components and a source of oxygen. Old-fashioned gunpowder is a mixture of combustible materials and a separate oxygen source. The combustible materials are charcoal and sulphur, and the oxygen source is nitre, potassium nitrate. Thus nitre had long been a strategically important material, as we saw when discussing Lavoisier's career in Chapter 3. In high explosives such as TNT (trinitrotoluene) and nitroglycerine (glyceryl trinitrate, the explosive ingredient of dynamite), the combustible matter and the oxygen source are combined together within the same molecule. The molecular framework is made up of carbon atoms linked together, with hydrogen atoms attached, and these are the combustible materials. The oxygen source is the nitro or nitrate chemical group, chemically bonded directly to this framework. When the explosive is detonated, rearrangement of atoms already present inside each molecule gives nitrogen, carbon dioxide, and water, all of them very stable substances, and a great deal of energy is released in the process. This sudden release of chemical energy is what drives the explosion, as the reaction propagates through the explosive material at more than the speed of sound. The products are formed extremely rapidly as gases (the water, of course, as steam) in a small space under high pressure, and the destructive shockwave of the explosion is caused by these gases expanding outwards from the high pressure region where they are produced.

Nitroglycerine itself is dangerously easy to detonate, but can be stabilised by kieselguhr, a kind of absorbent earth, making dynamite. Dynamite was invented by Alfred Nobel, a Swedish engineer and chemist, who became wealthy as a result of his work with this and other explosives. Nobel did not wish to be remembered only for inventions that contributed to the art of slaughter. Hence the prizes, awarded mainly by Swedish committees, except for the peace prize, which is administered by the Norwegian parliament.

All of these high explosives require nitric acid for their large-scale production.

In the first decade of the 20th century, Germany and Britain were clearly moving towards war. They were in an accelerating arms race, and the nations of Europe were building networks of mutually hostile alliances. War would need high explosives; high explosives, as explained above, would need nitric acid; nitric acid would need Chile saltpetre; and the British Navy could cut the Atlantic sea-lanes, starving the German munitions factories of this vital if unglamorous material.

The conversion of ammonia to nitric acid was already a well-established process, following a route that is still used today. So if you could make ammonia, you could make nitric acid. If you could make nitric acid, you could make high explosives, and that is what Haber, the patriotic German, was after.

As the first step in its conversion to nitric acid, the ammonia is reacted with air in the presence of a platinum catalyst, to give nitrogen monoxide (NO), also known as nitric oxide, and water. Ammonia will react with air without a catalyst, but the products of that reaction are just a useless mixture of nitrogen (N_2) and water. On the surface of the platinum catalyst, the reaction takes a totally different course. The ammonia molecule, NH_3, breaks down into separate atoms of nitrogen and hydrogen, no longer bonded to each other, but to the platinum surface. In the same way, molecules of oxygen (O_2) from the atmosphere dissociate on the surface to give separated oxygen atoms. The nitrogen atoms are the most tightly bonded on the surface, and tend to stick in their original positions, which stops them from pairing up to make N_2. Instead, they are picked up by oxygen atoms to make nitrogen monoxide, while other oxygen atoms react with hydrogen to make water. Nitrogen monoxide reacts further with oxygen to give the brown gas nitrogen dioxide, and eventually, if water is present, to give nitric acid. This can in turn be neutralised with ammonia, which is alkaline, to form ammonium nitrate.[3] The overall process of converting atmospheric nitrogen to a usable form is known as *nitrogen fixation*, and all forms of nitrogen other than N_2 are commonly referred to as "fixed nitrogen".

There are a number of other routes to fixed nitrogen, in addition to the direct ammonia synthesis. Traces of nitrogen monoxide are produced in small amounts whenever air is heated very strongly. That is why it is present in small amounts in automobile exhaust, and nitrogen dioxide produced by the reaction of exhaust gas with air is responsible for the brown colour of urban smog. We saw in the previous chapter how Cavendish had, as we now recognise, fixed nitrogen over a century earlier, by sparking nitrogen and oxygen together. On the modern industrial scale, the spark can be replaced by a continuous electric discharge. This can supply a temperature of 3000°C, and convert about 4% of the air to nitrogen monoxide. The nitrogen monoxide in turn can be converted as described above to nitric acid or, by reaction with alkali, to nitrate-containing salts. This was the basis of the Birkeland–Eyde process, operated at the Norsk Hydro plant in Norway (the first in the world to make synthetic potassium nitrate) using cheaply available local hydroelectricity. The giant German chemical company BASF must have thought that this was where the future lay, because they embarked on a major joint project with Norsk Hydro in 1907. Ammonia itself can also be produced by what is known as the cyanamide process. This involved a reaction cycle in which calcium oxide (made by heating limestone, and long used as quicklime in cement manufacture) was heated with coke in an electric furnace to make calcium carbide, which reacted with nitrogen at high temperature to make calcium cyanamide. Calcium cyanamide was reacted with water to make calcium carbonate and ammonia. The calcium carbonate could then be directly recycled, since on heating it gives up carbon dioxide and regenerates the starting material, quicklime.[4] Many plants producing ammonia by the cyanamide process were built in the decade starting in 1907. The process was still being worked as late as the beginning of the Second World War, despite the high energy input required for the first stage, and the word survived long beyond that as part of the name of the chemical company, American Cyanamid.

Ammonia could also be isolated as a byproduct in the production of coal gas, but the amounts available in this way were quite small.

As early as 1884, William Ramsay, discoverer with Rayleigh of the noble gases, had studied the decomposition of ammonia over hot copper to make nitrogen and hydrogen, the reaction that Rayleigh later used to make "chemical nitrogen" (as in the preceding chapter). One odd fact he noticed was that the reaction did not go entirely to completion, suggesting that it might be *reversible*.[5] Reversible reactions are a two-way street; they can go in either direction. In this case, *if* the decomposition of ammonia to nitrogen and hydrogen is reversible, that means that the recombination of hydrogen and nitrogen to make ammonia is also a possible reaction. In such reversible reactions, a dynamic equilibrium is established. Start with ammonia and, once you supply enough heat for the molecules to react, it decomposes to make nitrogen and hydrogen (which begin to react to make ammonia). Start with just nitrogen and hydrogen, and they will react to make ammonia (which will then begin to decompose). Either way, you will eventually end up with the same equilibrium mixture, in which formation and decomposition are happening at the same rate. The mixture is still reacting, but in both directions at the same time, and the two reactions cancel out overall.

In 1901, Le Châtelier in Paris reasoned (correctly, as discussed below) that applying high pressures to mixtures of hydrogen and nitrogen would increase the amount of ammonia present in the equilibrium mixture, but the work was terminated after a disastrous explosion. In 1904, Haber, with some commercial backing, turned his own attention to this reaction. Knowing that virtually all chemical reactions proceed more quickly at high temperature, he tried reacting nitrogen and hydrogen gases together over various catalysts at 1000°C. Some ammonia was formed, as he mentioned in his 1905 book, but, as he told his backers, too little to be of more than academic interest.

Why did such a simple seeming reaction present such enormous problems? To see this, we need to look a little bit more closely at what happens when molecules react. To turn a mixture of nitrogen and hydrogen into ammonia, you have to break apart the bond that holds together the two nitrogen atoms in the N_2 molecule, and the two hydrogen atoms in H_2, freeing up the hydrogen and nitrogen atoms to bond to each other and form ammonia, NH_3. These are all quite strong bonds. In fact the bond holding N_2 together is very nearly the strongest that there is.[6]

The energy to break these bonds comes from heat. In a gas, molecules are flying around bumping into each other at random, and the energy of their collisions with each other is directly proportional to the temperature.[7] Pretty well all chemical reactions go faster at higher temperatures, and this is especially true if the reaction involves the breaking of strong bonds. So this reaction is going to have to be carried out at high temperatures. But unfortunately, for this particular reversible reaction, at higher temperatures the equilibrium yield falls off very sharply.

Why should this be? Because the equilibrium depends on two opposing factors. The synthesis reaction gives out heat, and this is a factor that always makes a reaction more favourable. This is because heat is a very spread out form of energy, and according to the second law of thermodynamics those processes that increase this spreading, or *entropy* as it is called, are the ones most likely to occur. However, the reaction takes four molecules altogether, one of nitrogen and three of hydrogen, to make only two molecules of ammonia. This means a decrease in the total number of separate molecules flying around at random, so that the energy is less spread out, making the reaction less likely.

The principle describing such situations was formulated by the same Le Châtelier whose own work on this reaction came to such an unfortunate end. *Le Châtelier's principle* is an example of the cussedness of Nature. It says that if a system is at equilibrium, and we impose a change on the system, then the equilibrium will shift so as to partly undo the effect of the change. For example, if we have ice floating in a bowl of water, at equilibrium, the temperature will settle down at 0°C. If we try to warm up the system by supplying heat, the first thing that will happen is that ice will melt, pulling us back down to the original temperature. Exactly the same general principle applies in this particular reaction, where we have four molecules of reactant gas (one nitrogen and three hydrogens) reacting to give two molecules of gaseous product (ammonia). If we increase the pressure by compressing the gases, the equilibrium will tend to shift so as to moderate the increase. As Count Avogadro had first suggested a century earlier (see Chapter 4), the pressure is proportional to the total number of gas molecules,[8] so the equilibrium will shift in favour of the side of the equation where that number is smaller. It follows that high yields of product will be favoured by high pressures.

So far, so good, but the real problem is the effect of temperature on the equilibrium. Since the reaction releases heat, Le Châtelier's principle tells us that adding heat will reduce the yield.[9] In other words, higher temperature will shift the equilibrium away from the products back towards the reactants. This means that the reaction will give a good yield of product if and only if it can be made to take place at low enough temperature. Unfortunately, as explained above, breaking the very strong bond between the nitrogen atoms in the molecule N_2 requires copious energy in the form of heat, so there is a direct conflict between the need for the reaction to proceed at a useful rate, and the need for an acceptably high yield.

Heat of reaction, entropy, and equilibrium constants are part of the subject matter of *thermodynamics*, and the foremost practitioner of this subject in Germany at that time was Walther Nernst, whose "heat theorem", sometimes known as the "Third Law of Thermodynamics", is crucial when it comes to quantifying the entropy changes of reactions, and hence predicting how far any particular reaction will go at equilibrium. For better or worse, Haber and Nernst disliked each other intensely. Haber blamed Nernst for having blocked him from a professorship at Leipzig, while Nernst may have been jealous of the fact that Haber's publications on thermodynamics partly anticipated his own. In any case, the years from 1904 to 1907 saw both Nernst and Haber independently studying the reaction, initially as a purely scientific problem. Haber's yields were low, as we have seen, but Nernst's were even lower.

At a major conference,[10] Nernst presented his results, contrasted them with those that Haber had published, claimed that his own results were in agreement (as by implication Haber's were not) with predictions made using the "heat theorem", and attributed Haber's figures to experimental error. Haber, as one might imagine, replied at some length. With his new assistant Le Rossignol, he had repeated the work several times with similar results, used a variety of catalysts, and, crucially, had approached the equilibrium from both sides. Moreover, Nernst's own method of computation, when corrected (as Nernst had not done) for the different heat capacities of the gases, gave predictions in agreement with Haber rather than Nernst. Nernst stood his ground and, in a series of sharp exchanges, asserted that the prospects for ammonia synthesis were considerably poorer than had hitherto been assumed on the basis of "Haber's highly inaccurate numbers." At this stage, a tactful member of the audience intervened with an irrelevant question about the effect of electric fields on the reaction.

Haber now made a fresh estimate of the percent conversion to be expected over a range of temperatures and pressures, using the refinement he had made to Nernst's calculations. This led to the startling prediction that if one could lower the temperature to a few hundred degrees Celsius, rather than 1000°C, and carry out the reaction at a pressure of 200 atm, this would lead to commercially viable yields. With Le Rossignol, who had earlier worked in William Ramsay's laboratory at University College London, he adopted a two-pronged strategy. They would need to set about the design of apparatus that would stand such high temperatures and pressures, while embarking on a search for a catalyst that would allow the reaction to proceed at a reasonable rate, at as low a temperature as possible. The apparatus that they built had two crucial features. The hot gases leaving the chamber that contained the catalyst were cooled by exchanging heat with unreacted gas, and the ammonia that had been formed was removed as a liquid after cooling.[11] After extensive trials with a wide range of substances (the mechanism of catalysis being only poorly understood at this time) they discovered that the rare metal osmium would catalyse the reaction at 600°C, set up a small pilot plant that produced a steady stream of 100 cc (about a fifth of a pint) of liquid ammonia per hour, and filed a 1908 patent in both their names. At this point Haber felt able to advise BASF, with whom he already had a working relationship, that the time had come to move on to commercial production.

It was touch and go. BASF initially told Haber that they were not interested, in view of the technical problems, and it was only through the influence of his colleague Professor Engler, who sat on the BASF board, that the company was persuaded to change its mind, and back the project despite its ongoing scepticism.[12]

In 1912, BASF's rival, Hoechst, challenged the patent on the grounds that Nernst had already demonstrated the feasibility of the reaction, making Haber's contribution, in the sense relevant to patent law, "obvious". BASF responded by hiring Nernst himself as a consultant (at a fee roughly equal to $250,000 a year in today's money), to testify that this was not the case. After all, who better, and had not Nernst himself at the time pronounced commercialisation of the reaction to be impossible? BASF and Haber won their case, with costs, and Nernst and Haber were reconciled.

Heterogeneous Catalysts and How They Work

Why is the ammonia synthesis reaction so slow without a catalyst, and how does the metal catalyst work? When any chemical reaction takes place, new bonds are formed but also, at some stage, old bonds must be broken. To get from a mixture of nitrogen gas and hydrogen gas to ammonia, you have to break the very strong triple bond that holds the N_2 molecules together, as well as the hydrogen–hydrogen bond in H_2, which is one of the strongest of single bonds. (Why triple, and why should the H–H bond be particularly strong? See Chapter 12.) Metal catalysts give us a way to do this. They work because surface atoms have fewer neighbouring atoms than those in the stable interior. The interior atoms have the ideal number of neighbours for stable bonding, but the metal atoms at the surface have fewer than this ideal number, leaving them with unused bonding capacity. As a result, they can form bonds to atoms such as hydrogen or nitrogen. These atoms can then move from one surface location to another, forming new bonds to the catalyst as the old ones are left behind, until they meet up together and give products.

Very similar processes to these are involved in the conversion of ammonia to NO, and in the catalytic converter reaction that removes NO from vehicle exhausts.

Notice that metallic catalysis involves a whole sequence of separate steps. First of all, we have *adsorption* of the reactants, in which they stick on to the surface. Next, we have *dissociation*, in which the adsorbed molecules break down into smaller fragments or even, as in this case, to individual atoms. These atoms are bonded to the surface atoms, using the unsatisfied bonding ability always found at the surface atoms of a metal. The next step is *migration* of atoms over the surface, so that they can find new partners (this is referred to as *rearrangement*), and finally we have *desorption*, in which the product molecules leave the surface and re-enter the gas phase. All these separate steps are reversible, and taken together they lower the energy barrier to reaction in either direction. In this particular case, the key step is the breaking of the nitrogen–nitrogen and hydrogen–hydrogen bonds of the reactants, and the catalyst provides a low energy pathway by bonding to the separate N and H atoms as they form.

For the Haber process, the catalyst now used is chemically modified iron. This is unusual; it is more common to use metals like platinum as catalysts, and indeed we have met platinum as the catalyst in the next step of the process, the conversion of ammonia to nitrogen monoxide. In a good catalyst, the metal used must not be *too* reactive towards reagents or intermediates. If we tried to use iron in a car's catalytic converter, for example, the oxygen atoms would be too tightly held and never come off again; we would get clean exhaust for a while, but the iron "converter" would quickly turn into a pile of rust. Under oxidising conditions, therefore, we need a metal that will not bond too strongly to oxygen. Platinum is ideal; it forms moderately strong surface bonds to oxygen, but these are easily broken, and it does not go on to form a stable bulk oxide. Iron, if we tried to use it as a catalyst in a reaction involving oxygen, would simply corrode away.

Catalyst development continues to occupy a central place in molecular science. A catalyst cannot directly change the equilibrium yield of a reaction, because it speeds up the forward and reverse reactions in exactly the same ratio. However, there are three enormously important things that it can do. Firstly and most obviously, it can speed up the rate of reaction under given conditions. Secondly, as in the case of ammonia synthesis, it can make it possible to change the conditions under which a reaction is carried out, and in this way to change the equilibrium yield indirectly. Finally, a catalyst can change the path taken by a reaction, when more than one set of products as possible. We have seen an example of this in ammonia oxidation, where in the absence of the catalyst, we would have formation of nitrogen, rather than the desired product, nitric oxide. The most refined examples of this kind of control are at the heart of biology, where enzymatic catalysts steer the processes of metabolism along specific pathways, out of the enormous range open to the complex molecules on which life depends.

The Engineering Challenge

The optimum pressure for ammonia synthesis on the industrial scale depends on an economic cost–benefit calculation; the higher the pressure, the higher the yield, but very high pressure equipment requires sturdier, more expensive, plant, and in addition there is a higher running cost because of the larger amount of work that needs to be done to compress the gases.

Obtaining the high pressures is a matter of building sufficiently strong steel tubing. Getting the reaction to go at a satisfactory speed at low enough temperatures to give good yields is a major challenge. The amount of ammonia made is so great that even a marginal improvement in the process translates directly into major improvements in profits. If the Haber process could be made to work at a useful rate at room temperature, the yield would be almost 100%, and the major chemical companies still had teams of scientists working on new ways of improving his catalyst more than 50 years later.

In 1909, as we have seen, BASF took up the process, with a retainer that more than doubled Haber's salary. This was make or break time. Haber's results were certainly promising, but they had been obtained with osmium, and this metal is not cheap. (At the time of writing, it is trading at around 10% the price of gold.) Even more importantly, it is scarce, being obtained as a byproduct of the Canadian nickel mines, and this is of course a resource that Germany would not have access to in the event of war. The search for a better catalyst was led by Alwyn Mittasch at BASF, whose team conducted some 20,000 separate trials before eventually settling on the activated iron catalyst that is still used today. When it came to the vital task of moving from a laboratory scale demonstration to a full-scale chemical plant, Haber was most fortunate in having the services of Carl Bosch, also at that time a BASF employee. (Oddly enough, both Haber and Bosch had studied at Charlottenburg, although they would not have overlapped.) Bosch had trained as a chemist and metallurgist, although he may have drawn part of his practical inspiration from his family background; his father was a successful gas and plumbing supplier. He advised Haber that a 100 atmosphere pressure plant was feasible, and devised novel pumps, heat exchangers, and control and monitoring equipment; in fact Bosch is regarded as a crucial figure in the development of the entire modern chemical engineering industry.

One example will illustrate Bosch's ingenuity. The piping would have to be made of carbon steel, but it was found that this slowly became brittle, because at the high temperatures and pressures involved, hydrogen would react with the carbon in the steel to make methane, generating tiny flaws. Bosch's solution was to line the pipes with pure iron. Iron was not strong enough to take the pressure, but then it didn't have to. All it had to do was function as a barrier between the reacting gases, and the main body of the pipe.

A pilot plant was built in 1910, with production ramping up to 200,000 tonnes a year by 1918. In 1913, BASF sold its interests in the Birkeland–Eyde process that it had been operating with Norsk Hydro, reasoning correctly that the direct ammonia synthesis would make them worthless. Nonetheless, Norsk Hydro continued to operate the old process until 1929, when they bowed to the inevitable. The Norsk Hydro chemical facility continued to function in a variety of other roles, and was the site of the German efforts during World War II to obtain heavy water for their nuclear programme, successfully sabotaged by the Norwegian resistance.

Haber's Expanding Role

Over the period 1906–11, Haber built up his department in Karlsruhe to be among the best-equipped in the world, with optical and electrical apparatus as well as a balance room, library, and general laboratories. In 1912, the banker and philanthropist Leopold Koppel provided funds for a new research institute, the Kaiser Wilhelm Institute for physical chemistry and electrochemistry at Dahlem, Berlin, with Haber as director. Haber, with financial backing from Koppel, and on this occasion with Nernst's active collaboration, used this situation to lure Albert Einstein from Zurich to Berlin.

When war broke out in 1914, Haber volunteered for regular military service and was refused. The Ordnance Department recognised that he would be of enormously greater value in the laboratory than in the trenches, and set him to work on freezeproof gasoline for use on the Russian Front in winter, a matter of huge importance for troop mobility. This he accomplished by using additives that transformed the crystals that separated out at low temperature from hexagonal platelets that fouled the fuel line, to thin needles that were broken into small fragments by the pump.

The next task he was given was to create poisonous or irritating shrapnel, a task at which Nernst had worked and failed. He replied that it would make more sense to discharge the gas from cylinders, and this was the initial basis of his strategy.

On April 22, 1915, the Germans attacked with chlorine gas on the Western Front at Ypres; the attack left 5,000 dead and 15,000 injured, and was so successful that only lack of manpower limited the German advance. Chlorine, however, can be countered simply by using respirators, or gas masks. A more effective weapon is mustard gas, which causes skin blistering, and which was responsible for most of the chemical casualties in World War I. Haber claimed that the correct way to use gas was on a massive scale, and only if the war would be over within a year, since otherwise Germany's opponents would retaliate in kind (which they did) and would be helped by the prevailing Westerly wind over Northern Europe. Allied plans for 1919 — had the war continued — included driving tanks over the German trenches, spraying mustard gas.

Gas is not an easy weapon to use. The wind, one of the most difficult things to predict even today, has to be blowing in the right direction and at the right speed, ideally between around 4 and 9 mph. Small irregularities in the lie of the land, even in the lowlands of Flanders, will affect how the gas flows and settles. There is the serious problem of how the gas will affect one's own troops, if the wind changes direction, or even if they advance to occupy the territory abandoned by the enemy, as happened to the British when they in turn used chlorine gas in September at the Battle of Loos. The delivery of gas from cylinders is clearly only possible when the lines are very close together, and if it is to be delivered from shells, the charge that bursts the shell must be carefully controlled, otherwise the load will simply be dispersed over too large an area. One could also consider the morality of gas as compared with other weaponry, but that is not the main theme of our story.

Frau Clara Haber, meantime, was becoming increasingly unhappy. She had given up her own scientific career, and was worried by Fritz's overspending. She was also horrified by the use of war gas, which she regarded as a perversion of science, and tried to persuade Fritz to desist. The night before he was to leave for the Russian Front to install gas cylinders there, she shot herself with his pistol in the garden of their house. Haber did not hear the shot, and she was discovered by her son Hermann, then aged 13.

Fritz was to marry, two years later, a much younger woman, who was even less comfortable in the subordinate role he expected than Clara had been; that marriage ended in divorce after eight years. In his relationships with women, Haber compared himself to a butterfly-hunter, who thinks he has grasped his quarry, only to find that all he has in his hand is dust. Indeed, we might sympathise with the hunter in his disappointment, but what of the butterfly?

After the War, Haber wrote a letter to *Nature* claiming that it was only after war broke out that work on gas got under way, and that, as he put it, "In war men think otherwise than they do in peace, and many Germans during the stress of war may have adopted the English maxim, 'My country, right or wrong'".[13] It is indeed unusual for people to refuse requests from their governments in time of war. Michael Faraday refused to work on war gases during the Crimean War. Hubert Lamb, who went on to found the East Anglia Climate Research Unit, avoided working during World War II on the meteorology of poison gas spraying by moving from the UK to neutral Ireland,[14] but only one scientist, Joseph Rotblat, quit the Manhattan Project after it became apparent that Germany would collapse before any nuclear weapons became available.

In 1916 Haber became chief of new Chemical Warfare Service, with every detail of chemical offense, defense, supply and research under his supervision. This meant organising production for a four-year period in a country under a sea blockade of unprecedented intensity. The production of gas masks, for instance, involved the development of synthetic rubber, since Germany was cut off from supplies of the natural product. He was photographed with the Kaiser and was awarded the Iron Cross in both classes, the Order of Hohenzollern Swords, the Order of the Crown (Third Class), and various others.

Without Haber's contributions, the German war effort would have collapsed much more quickly, within three months according to the physicist Max Planck. As it was, Germany was able to sustain a far more damaging four-year-long conflict, receiving and inflicting horrible casualty levels, until its military and social collapse in 1918. Haber himself was officially labeled as a war criminal, and spent some time in Switzerland having escaped there under a false passport.

In 1919, he was awarded the Nobel Prize. There was no disputing the scientific importance of his accomplishment, but the award was highly controversial, because of his chemical weapons work. Nernst received his two years later.[15] Bosch was awarded the prize in 1931, for his contributions to high-pressure chemistry.

Consequences

Without Haber nitrogen, Germany would have had to sue for peace much earlier, and would have suffered far less in the process. As it was, 1918 saw revolution and collapse in Germany. The Kaiser abdicated, and the Peace of Versailles, based on the concept of national war guilt, imposed major territorial concessions and levied massive reparations, the accumulated interest on which was not finally paid off until October 2010.[16] The German economy collapsed under the strain, leading to massive unemployment and gross hyperinflation, with banknotes being overprinted in billions. Haber himself embarked on a typically ambitious plan to extricate Germany from its problems, by extracting gold from sea water, only to find that the gold content of the water had been greatly overestimated, and that the process could never be viable. In 1929, when the Great Depression spread throughout Western world, Germany was particularly vulnerable. The ensuing chaos paved the way for Hitler's Nazis.

In January 1933, Hitler became Chancellor of Germany. By the end of March, all civil liberties had been suspended. By April, Nazi student propaganda was libelling Jewish professors. Bosch attempted to form an association of non-Jewish scientists to protest government anti-Semitism, but the attempt petered out for lack of support. Bosch himself was later gradually relieved of his positions (though not before he had put in place a massive programme to convert coal to gasoline), took to drink, and died in April 1940.

Haber, recall, had been baptised, as had both his wives. He had his children baptised as well, occasionally went to church, and encouraged his Jewish-born friends to do likewise. None of this was relevant to the Nazis, who regarded Jewishness as a matter of "race".

In April 1933, a new law barred "non-Aryans", other than war veterans,[17] from employment in the civil service, including universities and research institutes. Although Haber himself was not directly affected, he was told by the Ministry of Art, Science, and Education that he could not continue to run his institute with its existing scientific staff. At the end of April, he decided that his position was intolerable and tendered his own resignation, effective October, giving as his reason that he needed to choose his scientific collaborators on the basis of their abilities, not the race of their grandmothers. Haber may have intended this interval for renegotiation of the employment policy. If so, he was sadly mistaken. His resignation was accepted, and his reasoning quoted by the Minister of Science as an example of a way of thinking that was quite unacceptable in the new Germany. Around this time, according to a surviving relative,[18] Haber sought out family members estranged since his conversion and said "I am a Jew again".

While Haber was technically a civil servant, his immediate superior, as president of the Kaiser Wilhelm Society, was the physicist Max Planck (Nobel Laureate 1918), creator of what we now call Planck's Constant, and one of the (somewhat unwilling) founders of quantum mechanics (see Chapters 10 and 11). Planck met Hitler himself in an effort to get Haber's resignation overturned; Hitler flew into such a rage that Planck had to leave the room.

By May 1933, the exodus of Jewish scientists from Germany was in full swing. Haber spent much of his time trying to find posts for those from his own laboratory, and discovered that foreign countries would usually only give them temporary positions because they did not want to damage the career prospects of their own people. Haber himself spent some time in Cambridge, at the invitation of W. J. Pope, who had been in charge of chemical warfare for the Allies. In August, he went to Spain to a conference, as a representative of the German Chemical Society, and then to Switzerland, hoping for relief from a deteriorating heart condition. There he met with the chemist and Zionist leader Chaim Weizmann. Weizmann's own wartime career had been an interesting one. A Russian Jew by birth, he had studied chemistry in Germany (including some time at Charlottenburg, albeit several years after Haber) and Switzerland. The war found him a naturalised British citizen, lecturing at Manchester, and later in charge of the Admiralty's laboratories. During this time, he developed a process for the large-scale microbial production of acetone, making him the father of industrial fermentation in much the same way that Bosch is the father of industrial high-pressure chemistry. In this case also, the motivation for the process was military. Acetone is used as a solvent, not only for nail varnish, but for guncotton.

In 1933, Weizmann was among the most prominent leaders of the Zionist movement, and was in a position to offer Haber the Chair of

Physical Chemistry in the fledgling Hebrew University of Jerusalem. As the correspondence between them makes clear,[19] Haber showed enthusiasm for the idea, brokered Weizmann's hiring of his junior colleague Ladislaus Farkas (by then in Cambridge), discussed practical details even down to the level of provision of kitchenware, and urged Weizmann to embark on an ambitious hiring programme, to take advantage of the gap being created by Germany's inevitable imminent scientific decline. In January 1934, however, Haber had to inform Weizmann that he had been warned that the journey would be too strenuous for his deteriorating heart, and within the month he was dead.

On the first anniversary of Haber's death, Max Planck organised a memorial ceremony at the Kaiser Wilhelm Society conference centre. It was an extraordinary occasion. The Nazis had ordered Haber's own former colleagues to stay away, but the ceremony took place nonetheless and was attended by their wives. Planck himself delivered one eulogy of his own, and one on behalf of his absent colleague Karl Friedrich Bonhoeffer. Karl Friedrich was the brother of the pastor Dietrich Bonhoeffer who was later executed for his links to the 1944 plot to assassinate Hitler. Planck's address reminded the audience that without Haber's contribution, Germany would have collapsed in the first three months of World War I. Max von Laue (one of the pioneers of x-ray crystallography; Nobel prize 1914) had written an obituary for the journal *Naturwissenschaften* with rather different emphasis, describing Haber's discovery as yielding "bread from air". Planck and Laue were the only two German scientists of whom Einstein (who had left Germany in 1932) later spoke with any respect.

From 1939–45, Hitler led Germany to a further disastrous war, leading to more territorial losses. And so German Breslau, Haber's birthplace, became Polish Wrocław. Ladislaus Farkas had a brief but distinguished career at the Hebrew University, was involved during World War II in the local production of chemicals for the Allies' Middle East Supply Centre, and was killed in a plane crash in 1948, aged 44. Weizmann became the first President of the State of Israel, and the research institute at Rehovot that he helped found is now named after him. The network of Kaiser Wilhelm Institutes were renamed the Max Planck Institutes, except for the one at Dahlem, now named after Fritz Haber.

Chemical Weapons Today

Chemical weapons were used by Mussolini's Italy in Ethiopia, and by the Japanese in China, and it was widely expected that chemical weapons would be used in Europe in World War II. The United Kingdom government built a major installation in Wales for the production of mustard gas, and issued the population with gas masks. The children's masks were made of pink rubber, and stylised to look like Mickey Mouse (actually, I was rather fond of mine). However, the expected attacks never took place, perhaps because of the practical difficulties of deployment on rapidly moving battlefronts or from all but low-flying aircraft.

The use of chemical weapons is now totally banned by international agreement. In the US, their manufacture is forbidden by federal law, as well as by an international treaty — the Chemical Weapons Convention — which the US Senate finally ratified in 1997. As required by the Convention, the US is destroying its existing stocks, an expensive and hazardous procedure for which a projected completion date of 2021 now looks optimistic.[20] The "yellow rain" alleged to be a form of chemical warfare directed against US troops in Vietnam proved to be nothing more than bee droppings, but was used at the time to help persuade the US to build up its own chemical weapons stockpiles, which are now, as already mentioned, being destroyed at great expense. While chlorine and phosgene are obsolete as war gases, mustard gas was used by the Iraqis in their war with Iran and in suppressing uprisings, as were *nerve gases*, which block reactive centres in nerve cells; these are very effective weapons because they work by skin absorption, as well as by inhalation, so that gas masks offer poor protection. During the 1991 Gulf War the Iraqis stockpiled nerve gases, but did not in fact use them. However, five years after the event, the Pentagon admitted that large numbers of US troops may have been accidentally exposed to low levels of these gases when an Iraqi munitions dump was blown up at the end of that war. After 1991, the Iraqi stockpiles were destroyed under United Nations supervision. Claims that the chemical weapons programme had been revived were among the reasons given by the US and UK governments for the 2003 invasion of Iraq, but no evidence for such a revived programme was ever found.

In November 2009, over six years after the invasion, it emerged[21] that UK ministers were repeatedly warned of the unreliability of their intelligence on Iraq, that days before the invasion they were told that Iraq's chemical weapons "might have remained disassembled and Saddam hadn't yet ordered their assembly," and that "Iraq might lack warheads capable of effective dispersal of agents." Effective dispersal, as discussed earlier, is of course crucial to the effective use of such weapons.

Fertilisers, Population, and the Haber Process

We can never be sure of what our actions will lead to. Haber sought to serve his country; his actions, as we have seen, helped destroy his country, his family, and himself. Yet now, half the world's population depend on his discoveries for their very survival, because of the vital role in food production of synthetic fertilisers using Haber process ammonia. Proteins are built out of *amino acids*, which are molecules that contain both the $-NH_2$ (*amine*) grouping and the $-COOH$ (*carboxylic acid*) grouping (or more precisely, the $-NH_3^+$ and $-COO^-$ groupings). Ordinary plants cannot build amino acids using the nitrogen in the atmosphere, and have to rely on the nitrogen fixed by bacteria in the soil or applied from outside as fertiliser.

From ancient times until less than two hundred years ago, hunger was regarded as part of the normal lot of mankind. Thomas Malthus, in his *Essay on Population* (1798), which helped inspire Darwin, argued that famine would become increasingly common, since populations would grow until limited by lack of food. In the 1840s, it seemed as if Malthus' prediction might be coming true, with Ireland, Holland and Belgium, and parts of Scotland experiencing starvation when the potato crop failed. However, more wide-spread famine was averted by the use of artificial fertilisers, such as Chile saltpetre, together with other improvements in agriculture, and then from around 1870 on by the opening up of the great grainlands of the Americas.

Something very similar occurred a century later. 1947 was a year of near-famine in Europe; the resulting tensions had a major influence on the development of the Cold War between the Western and Soviet blocs. At the same time, antibiotics and safe drinking water (the latter based largely on the use of chlorine) were becoming available on a large scale, so that population was set to rise, if only because people were living longer.

Disaster was averted by the *Green Revolution*, which was the development, by systematic breeding, of new, much higher-yielding, strains of grain crops, especially wheat and rice. These strains produce shorter stalks and, when grown on sufficiently fertile ground, more abundant ears and more harvests each year than the older varieties. To do this, they require higher levels of fertiliser than could be provided by traditional techniques. Grains such as wheat, rice, corn, oats and barley are all grasses, and grasses require large amounts of nitrogen to support rapid growth.

Today, only about half the fixed nitrogen used in world agriculture is provided by soil bacteria. The rest, without which the great growth in crop yields would not have been possible, comes from products of the Haber process. Ammonium salts are excellent nitrogen-providing fertilisers, as are nitrates, and their combination in ammonium nitrate is widely used to provide fixed nitrogen. The Green Revolution banished the spectre of global famine for at least one generation, and countries such as India and China, which before the Green Revolution were regularly on the verge of famine, became net food exporters. This is obviously a triumph of applied biology, but its success depends just as much on chemistry. To quote the agriculturalist Norman Borlaug, who earned a well-deserved Nobel Peace Prize for his share in the plant breeding work, "If the high-yielding dwarf wheat and rice varieties are the catalysts that have ignited the Green Revolution, then chemical fertilizer is the fuel that has powered its forward thrust."

No change as dramatic as this occurs without costs. Small farmers near the subsistence level, for instance, often cannot afford seed and fertiliser, and may be forced off the land by their wealthier and more efficient neighbours, with great damage to the social structure. This explains why international aid agencies are now shifting attention from large-scale assistance for governments to small-scale assistance for family farmers, helping them form collectives that can make the necessary capital investments.

Further Prospects

Despite intense research efforts and ongoing studies,[22] there have been no fundamental improvements to the Haber-Bosch process in a century. The root cause of this is the extreme stability of the N_2 molecule. There are a few examples of transition metal compounds that will interact strongly enough with nitrogen to make it reactive towards hydrogen at room temperature, but these are themselves irreversibly converted to metal hydrides in the process.[23]

Between 1947 and the mid-1980s, world food supply outgrew world population. Since 1984, however, this has not been the case, and further demands on land use are bound to arise as India and China continue to industrialise. Industrial cities tend to grow on reasonably level sites with good communications and water; this generally means in fertile river valleys, putting the cities into direct competition with agriculture. Recently, world grain prices have been rising steeply, an ominous sign, and the current generation of biofuels are in direct competition with food. The United Nations Food and Agricultural Organisation reports[24] that global food security has been deteriorating since 1996, with more than a billion people now hungry or undernourished, a situation made worse by soaring food prices, followed by the world economic crisis. At the time of writing, China is undergoing severe food price inflation, driven by loss of arable land. Food prices are rising steeply in the rest of the world as well, hitting the poorest hardest, and bringing unprecedented political challenges even to well entrenched dictatorships.

It is not clear at the moment where the next large increase in world food production could come from, or how the balance between population and food supply will be maintained, especially in the face of global warming (Chapter 15). To quote Norman Borlaug again,[25] in a passionate plea for the use of all available resources (including genetic manipulation), "It took some 10,000 years to expand food production to the current level of about 5 billion tons per year. By 2025, we will have to nearly double current production again … These problems will not vanish by themselves. Unless they are addressed in a forthright manner future solutions will be more difficult to achieve."

The nitrogen and hydrogen for the present-day version of the Haber process come from carefully controlled reactions between natural gas, water, and air. The process could readily be adapted to use coal instead of natural gas, and the world's reserves of natural gas and coal are large. Thus in the near future ammonia production is less likely than the availability of land and water to be a limiting factor on the world's food supply. Ammonia production from fossil fuels contributes to the increase in atmospheric CO_2, and hence to global warming. This is not trivial, since something like 1% of the world's entire energy use is connected with the Haber process. Even more serious is the formation of nitrous oxide, laughing gas, from nitrate fertiliser. This is a leading ozone depleting gas, and a powerful greenhouse gas in its own right. In the long run, the hydrogen required could be obtained from the electrolysis of water, but such is the efficiency of the conversion of fossil fuels to hydrogen for the process, that it seems likely that it will continue to be worked in the same way until these fuels are exhausted, even after their use in power plants is phased out, as it surely will be.

Conclusions

How should we judge Fritz Haber, who set out to save his country and helped destroy it, and whose activities made possible what was, in terms of purely military casualties, the bloodiest war in the whole of human history? Should we excuse him, as a man of his times, in much the same way that we might excuse Washington and Jefferson for being slave-owners? How should we evaluate the morality of his work on war gases, and how should we weigh this against the enormous effects of the Green Revolution made possible by his work?

I asked these questions at a talk I once gave, and a historian present gave a memorable answer. She said we have no business to sit in moral judgement on Haber at all. We are not in his situation, and he is beyond our reach. We have our own dilemmas, and our actions will have their own ironic consequences. We should not attempt to judge him, but rather to learn from his fate, for moral choices are inescapable, and history does not stop.

To Sum Up

High yields of grain crops require the use of nitrogen-containing fertilis-ers. In the early 20th century, the main source of these was Chile saltpetre, but supplies were limited, and vulnerable to interruption (especially from Germany's point of view) in the event of war. Such fertilisers could, how-ever, be prepared from ammonia, and thus the production of ammonia from nitrogen and hydrogen was a central problem in chemistry. The reaction proceeds very slowly, except at high temperatures, but higher temperatures lead to lower yields, for fundamental thermodynamic reasons. Haber and his collaborators solved the problem by using higher pressures, thus increasing the yield, and a catalyst, which means that the reaction can be run at lower temperature.

All this occurred in the run up to World War I. Haber was a fiercely patriotic German, and his main motivation came from the use of nitric acid, prepared from ammonia, in the production of high explosives. (His other contributions included the production of war gases.) The ammonia synthesis made it possible for Germany to fight a much longer and more destructive war. The resulting defeat and social collapse were to pave the way for the rise of Hitler, and Haber, a Jew by descent, died of heart failure in Switzerland.

7 The Ozone Hole Story — A Mystery with Three Suspects

A crime has been committed, and there are three suspects. The first is a well-known repeat offender, but will turn out — on this occasion at least — to be working for the authorities. The second is an apparently innocuous and highly valued citizen, with friends in high places. The third has a strong alibi, but faces politically motivated prosecution, based on circumstantial evidence. The mystery is finally solved using spy technology, and the whistle-blowers richly rewarded.

The Victim

Ozone is a high-energy form of the element oxygen, containing three atoms of oxygen. The usual form of oxygen contains just two.

Ozone is less stable than oxygen, and at equilibrium the amount of ozone expected to be present is completely negligible. However, the stratosphere is not at equilibrium because it is being bombarded with ultraviolet energy direct from the Sun, leading to the formation of an *ozone layer* in the stratosphere, mainly between 15 km and 30 km above the Earth's surface. The amount of ozone present is very small, so little that if you could separate it from the other gases present, and bring it to room temperature and pressure, it would only form a layer around 3 millimeters thick. Nonetheless, this layer plays a vital role in making life on Earth possible.

112

Everything of real interest happens far from equilibrium. The entire drama of life is possible only because the Earth, at an average temperature of some 15°C, or 288°C (518°F) above absolute zero[1] is far from equilibrium with the sunlight generated by a hot body glowing with a surface temperature of 6000°C. In the case of the ozone layer, this light is disturbing the equilibrium directly.

As we saw in Chapter 4, ordinary oxygen exists in the form of O_2 molecules, containing two oxygen atoms joined together. In the ozone layer, which lies in the lower stratosphere, high energy ultraviolet light from the Sun splits some small fraction of the oxygen molecules into their separate atoms. These atoms can then attach themselves to a molecule of ordinary oxygen, O_2, to make ozone, O_3. Eventually, however, this ozone molecule will meet another oxygen atom, and react with it to give two ordinary oxygen molecules.[2]

The rate of ozone destruction depends on the amount already present, although the rate of formation does not. Imagine that there is no ozone to start with; then ozone can only be made and not yet destroyed, since there is none there to be destroyed. However, if we could start with a more than steady state concentration of ozone, then the rate at which it was being destroyed would be greater than the rate at which it was being formed, and there would be a net fall in its concentration. There will be one particular abundance of ozone at which the formation and destruction reactions go at just the same rate, and that is the steady state abundance. So the balance between formation (by oxygen atoms reacting with O_2 molecules) and destruction (by oxygen atoms reacting with ozone itself) regulates the amount of ozone in the stratosphere.[3]

Or so it seemed.

As to why this matters, ordinary oxygen, O_2, absorbs light of wavelength less than 190 nm (1 nm, or nanometre, is one billionth of a metre). Ozone in the stratosphere absorbs UV light of wavelength between 190 nm and 240 nm, which would otherwise break bonds in living tissue, causing mutations (by breaking bonds in DNA), burns, and cross-linking of proteins. All these things happen to some extent even with the ozone layer present, under the influence of light between 240 and 280 nm. We can get sunburnt, the Sun causes aging of skin and loss of its elasticity, and more exposure to Sun increases our risk of melanomas, or skin cancers. However, light of wavelength shorter than

240 nm is even more damaging, and life on land only became possible after enough oxygen had built up to produce a protective layer of ozone.

The Crime

Around 1960, the amount of ozone over Antarctica began to decline, especially during the southern spring month of October. At first, it was difficult to be sure that the effect was real, but the evidence grew stronger as time went by. By 1985, the October Antarctic ozone was below 60% of its 1950s level, and by 1992 it was less than half.[4] The effect was not confined to the Southern Hemisphere or to the polar regions. The amount of light in the region causing sunburn increased in Torouto by 5.3% a year in summer and 1.9% a year in winter between 1989 and 1993.[5] Slightly longer wavelength light, which is not absorbed by ozone, showed no change during that time, ruling out changes in such things as solar activity. Separate measurements of the amount of ozone showed that it was indeed decreasing, and that this would account for the extra ultraviolet light.

Could this damage to the ozone layer be the result of human activity? There were no natural changes between 1960 and 1990 to account for the ozone loss, but there were two changes in human behaviour over this period that could be responsible; these provide our first two suspects. The third suspect is a natural process, framed on circumstantial evidence by those who could not or would not believe that human activities might affect the entire planet. It will turn out that one suspect has an alibi, one has been caught with a smoking gun, and one is now cooperating with the authorities.

Does it matter? Yes; thinning of ozone is detectable over fairly large areas. It causes real damage by the time you get to the south end of South America, with a 50% increase in the skin cancer rate in Punta Arenas, the world's southernmost city, between 1987 and 2000.[6]

The Investigation

The forensic tools for this investigation were developed in the decades immediately following World War II, and owe a great deal to the massive increase in public funding for science at that time. This was driven by appreciation of the crucial role science had played in the Allied victory, by a growing awareness in political circles of the economic benefits of scientific discovery, and, it must be said, by Cold War rivalries.

The most obvious requirement was a good way of studying extremely fast chemical reactions. Oxygen atoms, which play such a crucial role in the formation and destruction of ozone, are highly reactive, and do not last very long, but without knowing just how quickly they react, it was not possible to test the suggested mechanism.

Several techniques were developed to study fast reaction *kinetics*, all of them depending on using a short pulse of energy to disturb the equilibrium in a system, and studying the rate of disappearance of the reactive fragments produced by the pulse. Chief among these is *flash photolysis*, developed by Professor Norrish at Cambridge and his graduate student, George Porter (later Baron Porter). Norrish was one of the leading photochemists of his day, and Porter had been, during World War II, involved in the use of radar. From this he had derived a training in electronics, and an awareness of the additional information that could be obtained by using pulsed, rather than continuous, signals. The flash photolysis technique, as the name suggests, involves subjecting reagents to a brief, intense pulse of light and studying the rate of disappearance of the reactive species generated. Norrish and Porter shared the Nobel Prize for chemistry in 1967, along with Manfred Eigen who had developed other fast reaction techniques. The subsequent development of flash photolysis has depended on techniques for producing ever-shorter light pulses from lasers, until these pulses are comparable with the timescale of a singular molecular vibration, and, in the words of Ahmed Zewail (Nobel prize, 1999), the race against time has been won.

The investigation also required techniques for studying the actual chemical composition of the upper atmosphere, including methods for the detection of trace components. Motivation for this came from the need for improved weather forecasts, and from the growing importance of both military and civil aviation, with the development of passenger jet aircraft routinely flying at about 30,000 feet (around 10 km), and the U2 spy aircraft, flying in the stratosphere at 70,000 feet (more than 21 km). These were operated by the CIA, causing much embarrassment when, in 1960, one of them was shot down over Russia. (Although this incident caused Eisenhower much embarrassment in his relationships with

Khrushchev, such flights, like the satellite surveillance that succeeded them, merit rational approval, for calming exaggerated fears of Soviet intentions.) U2 aircraft themselves were used to collect samples from the atmosphere, as were high flying balloons. Such balloons could also be equipped with infrared spectrometers, which measured which wavelengths of light from the Sun were absorbed by the atmosphere, and by conducting such experiments while the Sun was setting it was possible to see how the intensity of absorption changed at different levels. The infrared absorption spectrum of a substance acts as a fingerprint, and spectra could be matched against those obtained for known materials in the laboratory. Rocketry also contributed. Ozone concentrations were measured at heights up to 70 km using captured V2 rockets. In 1956, Fred Singer suggested refining the spectroscopic techniques for measuring ozone by looking for its absorption spectrum in sunlight bounced off satellites (we will meet Fred Singer again, in a rather more sinister role, when we discuss global warming). Developments in gas chromatography, associated in particular with the name of James Lovelock of Gaia Hypothesis fame, made possible the detection and measurement of trace materials down to the parts per trillion level.

The third major development underlying the investigation was the advent of scientific computing. This again grew directly out of wartime developments, such as Alan Turing's work on decoding the Enigma machines used by the Germans to encrypt communication with submarines. Because of the complexity of weather systems, meteorology departments were early users of computers, and one of the leading investigators, Paul Crutzen, actually entered academic life through the back door in 1959 as a computer programmer for the Department of Meteorology of Stockholm University.[7]

Contrary to popular misconception, the use of computers enormously increases the reliability of scientific predictions. In quantitative science, the fundamental laws of nature are often expressed in terms of rates of change; for example, Newton's Second Law states that the rate of change of velocity of any object is proportional to the force applied to it. But it isn't the rate of change that usually interests us. We want to know how fast an object will be moving after a certain time. That means carrying out a mathematical operation called *integration*, in order to find out the effects of the change accumulated (integrated) over that time. Indeed, since velocity is rate of change of position, we will need to carry out a second integration in order to find out where the object ends up. Using integration to calculate the outcomes of a law that is expressed in terms of rates of change was, in fact, the dominant procedure of the physical sciences for almost 300 years.

Yet this method is extremely limited. For example, using Newton's laws of motion and of gravity, we can use integration to make mathematically exact predictions for the Earth–Moon system considered on its own, or for the Sun–Earth system considered on its own, but there *is* no exact way of solving the real problem, of predicting the behaviour of the three-body Sun–Earth–Moon system in its entirety. Before the advent of computers, the best we could do was to consider the exact solution for part of the problem, and then use various approximations to estimate how that exact solution would be perturbed by the other components. Now, we can apply the laws of motion to objects within a computer simulation, allow them to move a short distance, then recalculate the forces acting on them and allow them to move again. With enough computing power, we can make the intervals as short as we like, until we are satisfied that further shortening will make no appreciable difference.

Rather similar reasoning applies to interlocking sets of chemical reactions, where we have several different kinds of molecule each playing a role, and some molecules playing more than one. We can, with any luck, study the rate of any individual reaction on its own, by looking at how quickly two or perhaps three of these kinds of molecule react with each other when mixed together in the laboratory. When the product of one reaction acts as a catalyst for others, it becomes impossible to carry out the integrations that will tell us where the system winds up, but with the computer we can still make predictions by repeatedly allowing the system to change over very short periods of time, and after each step re-calculating what the rates will be under the new conditions.

Not only does this enable us to solve otherwise intractable problems, but it enables us to find out whether or not our solutions are *robust*. We can run the calculation using slightly different assumed conditions or rate laws, and see how much difference that makes to the predicted outcome. The more robust the predictions are to such change, the more confidence we can have in them. In the extreme case, where even a very small change in initial conditions is amplified and eventually changes the total output, we have the *chaos* discussed in Chapter 10. Precisely this situation can arise in the gravitational three body problem, in weather prediction (which is why long-term weather prediction is impossible), or in certain chemical systems, although, fortunately, not the particular ones that we need to consider here.

So using the computer enables us to do two rather different but related things. It enables us to carry out computations on complex systems, and then to test how vulnerable our conclusions are to small changes in our assumptions. The cynical saying "Garbage in, garbage out" certainly applies to computing, in exactly the same way that it applies to any kind of reasoning process whatsoever, but the computer gives us a much better chance of identifying both the garbage itself and the assumptions that gave rise to it.

We can now allow our whistleblowers to get to work.

Suspect #1: Nitrogen Oxides

We have seen in qualitative terms how the concentration of ozone will reach a steady state, under the influence of oxygen atoms involved both in its creation, and in its destruction. Using flash photolysis, we can generate oxygen atoms in the laboratory, follow their rate of reaction, and extrapolate to the conditions in the stratosphere. Surprisingly, it turns out that the rate of disappearance of ozone calculated in this way is smaller than the observed rate by a factor of two. Something else is going on. What?

Paul Crutzen, whom we met as he joined the computing staff of Stockholm University's meteorology department, collected a Masters degree in science by studying mathematics, mathematical statistics, and meteorology (no physics or chemistry; he had a full-time job and the lab courses would have taken up too much time). The department was a world leader in the study of atmospheric chemistry, and one of the first to study the problem of acid rain. Crutzen wanted to study the fundamental science behind atmospheric processes, had been given the job of helping a colleague model ozone distribution, and chose the formation and destruction of ozone in the stratosphere as the problem for his Doctoral dissertation.

Water is split in the stratosphere by ultraviolet light into hydrogen atoms and OH (hydroxyl) groups. OH is an example of what chemists call a *free radical*, nothing to do with libertarian politics but a group of atoms (a "radical") that normally occurs as part of a molecule, wandering around on its own. Such species are highly chemically reactive, and it had been suggested that it was OH that was destroying ozone, by removing an oxygen atom in a catalytic cycle.[8] Crutzen showed that this process didn't fit the facts about the distribution of ozone in the stratosphere. However, as he showed, OH is indeed an important player in other aspects of atmospheric chemistry, including the destruction of methane and other hydrocarbons, which stops them from reaching the stratosphere. He therefore suggested that there might, on the other hand, be a role for oxides of nitrogen.

The chance to explore this idea came in 1969. He was in the Department of Atmospheric Physics at Oxford, as a European Space Research Organisation postdoctoral fellow, when the head of the research group, John Houghton, showed him a spectrum of the Earth's atmosphere taken from a balloon, and mentioned that it might show bands due to nitric acid, HNO_3. (Houghton later became chair of the Intergovernmental Panel on Climate Change, a topic we discuss in Chapter 15). Yes, indeed, the relevant bands were there. But under the conditions present in the upper atmosphere, nitric acid interconverts with other compounds, including in particular NO. This was interesting. NO is just the kind of molecule that can remove a single oxygen atom from ozone, converting it to normal oxygen, and Crutzen proposed the existence of a simple catalytic cycle based on this reaction.

What happens is that in the stratosphere, NO reacts with ozone (O_3) to give normal oxygen (O_2), and NO_2, the choking brown gas found in urban smog. In addition, NO itself can trap oxygen atoms to give NO_2. Finally, NO_2 itself can be removed in one of two different ways. It can absorb light (remember that it's coloured), splitting apart to give NO and an oxygen atom, which will subsequently react with O_2 to regenerate ozone. 95% of the NO_2 formed reacts in this way, in which case no harm is done. But in the stratosphere, where oxygen atoms are relatively abundant, 5% of the NO_2 reacts with these oxygen atoms to give NO and O_2, thus removing "odd oxygen" (the name given to interconverting ozone and oxygen atoms) from the system. (At ground level, NO reacts by a more roundabout route with atmospheric oxygen to give the choking brown gas NO_2 as part of the process of smog formation.)[9]

So where could the NO in the stratosphere be coming from? It is known that traces of NO are produced whenever the nitrogen and oxygen in the air are strongly heated or sparked together. This is the same reaction that was studied by Henry Cavendish (Chapter 5), and that forms the basis of the Birkeland–Eyde process for nitric acid (Chapter 6). This process also occurs naturally in lightning, and was the source of the reactive nitrogen required by the earliest living things on Earth. It occurs, unwanted, in the internal combustion engine, and, inevitably, in jet engines.

Commercial jet travel began in 1952, and has grown rapidly since that time. Ordinary jets cruise at a height of around six miles, close to the lower limit of the ozone layer. Thus NO from jet engines is our first suspect; it is formed in more or less the right place, and its increased production in the third quarter of the 20th century coincided with the observed depletion of ozone.

Further work showed that there is indeed NO present in the stratosphere, and in far higher amounts than could have arisen by upward mixing of the NO generated by aircraft. It actually arises rather indirectly, from nitrous oxide, N_2O. Nitrous oxide is the "laughing gas" of dental anaesthesia, with a history of recreational use dating back to the Royal Society in the 18th century; Davy, whom we met briefly in Chapter 4, held laughing gas parties for the poets Coleridge and Southey. It has a number of other uses, including the aeration of whipped cream. It is formed in Nature by bacterial action on nitrates, and the amount in the atmosphere has been slowly increasing as a result of human activity, in particular the use of nitrate fertilisers, as described in Chapter 6. It also happens to be a greenhouse gas; more on this in Chapter 15.

Measure the reaction rates, do the calculations (using a computer of course), and you will find that the NO from nitrous oxide is responsible for roughly as much ozone destruction as the reaction between ozone and oxygen atoms. So NO is indeed of enormous importance here. But the amount of NO in the stratosphere from ordinary jet travel is small compared with the amount there already. Additionally, very few jets fly over the Antarctic, which is where the largest hole appears. So we can regard NO in general as guilty of ozone destruction, but the added NO from jet travel is a minor player, and certainly not responsible for the ozone holes over the polar regions.

Things might have been very different, and far worse. In the late 1960s, plans were widespread to build fleets of supersonic aircraft (generally known as supersonic transports, or SSTs). However, it was clear that these would be injecting their NO directly into the ozone layer itself, and calculation showed that with large SST fleets, there would be unacceptable thinning of the ozone worldwide. That is one reason why the US, in particular, did not go ahead with these plans. The other reasons were commercial. The sonic boom created by SSTs was so loud that it could only be tolerated over open water, greatly reducing the range of routes where they could be used. Widebodied aircraft were coming into service and these were much more economical in every way, including using much less fuel per passenger mile. Trans-Atlantic air travel (the obvious candidate for SSTs, and indeed the only purpose for which they were ever used) was also becoming more comfortable. The economic aspect grew in importance over time, as fuel costs rose and they seemed (and seem) likely to rise even further in the future. Besides, why spend a fortune flying Concorde, in order to save a few hours for reading, watching a film, or having a nice dinner, when as a first-class subsonic passenger you can read, watch films, and have a very nice dinner on the plane, for half the price?

Apart from a short lived Soviet adventure, the only commercial aircraft that ever became operational was the Anglo-French Concorde, which saw service from 1976 until 2003. Even at the speeds it was allowed to reach over land, it generated enormous noise and vibration. Houses beneath its flight path had to be equipped with sound-absorbing multiple glazing, and on the one occasion when I happened to be outdoors when it passed overhead, I could feel my teeth rattling in my head. Never have so many been so inconvenienced for the benefit of so few, and I was not sorry to see it go.

We were lucky. If, by chance, the relative rates of development of aeronautical engineering, and atmospheric chemistry, had differed by only a few years, we would have learnt about the effects of NO on ozone the hard way, through increases in sunburn, skin cancer, and crop damage.

This looks like a good example of the textbook scientific method in action. Problem: what is causing the ozone hole? Hypothesis: that NO is responsible. Experiment: in this case *experimental modelling*. Direct experiment is not possible, since we could hardly set about increasing the amount of NO in the stratosphere, nor would we risk doing so if we could. So we use indirect experimental evidence, by measuring the rates of the different possible reactions in the laboratory, collecting information about conditions in the stratosphere, and combining all these data to calculate the rates that would occur there. Testing of hypothesis: the rate of the process is not sufficient to explain the hole. Modification of hypothesis in the light of new information.

Actually, it is nothing of the sort. It is a meandering tale of circumstance and conjecture, of individuals happening through accidents of time and place to get interested in particular problems, of intellectual puzzles turning out to be of unforeseen practical relevance, and of continual dialogue between theory and experiment, and between theoreticians and experimentalists. It is a good example of how science actually happens.

Suspect #2: Chlorofluorocarbons (CFCs), e.g. CF_2Cl_2

These are the most innocuous compounds imaginable; that's the problem! They are non-toxic, non-flammable, completely non-reactive in the environment, and don't absorb any light that reaches the Earth's surface. Because CFCs are so unreactive, it was believed that there were absolutely no dangers associated with their use, so that they were formerly used (no more) to pressurise aerosol cans, to puff up Styrofoam, and even (since they happen to be good solvents for oxygen) in synthetic blood plasma substitute. They were also used as solvents, to clean circuit boards in the microelectronics industry, and as the working fluids of air conditioners and refrigerators.

So where does a CFC molecule go to die? To the region around the top of the ozone layer, where the UV energy is enough to chop apart the bonds in the molecule, giving Cl atoms. Each Cl atom then catalyses the decomposition of ozone. Chlorine atoms react with ozone to make chlorine monoxide, which can then react with an oxygen atom to form O_2 and regenerate the original chlorine atom.[10] The Cl atom can eventually react with methane (from biological activity) to make HCl and the $CH_3\cdot$ radical (which is eventually oxidised to CO_2), but methane does not make it into the stratosphere, because (as explained above) before it can get there it reacts with photochemically produced OH. Chlorine atoms are removed by other means, as we shall see later, but before this happens each one is estimated to destroy 100,000 molecules of ozone.

Suspect #3: Volcanic Emissions

Volcanic emissions are our Suspect #3. They inject far more chlorine-containing material into the lower atmosphere than all other sources put together, making them an obvious scapegoat. It is, however, rather difficult to blame volcanoes for a process that only started after 1950. They also have a good alibi. Their chlorine is in the form of hydrogen chloride, which is extremely soluble in water and is washed out by low-lying rain clouds long before it can reach the scene of the crime.

This is all very well, but in 1991, Mount Pinatubo in the Philippines erupted, emitting large amounts of hydrogen chloride and dust, and causing cool summers worldwide for two years. During these years, the ozone hole was deeper than usual; why? The alibi is at best negative evidence: how can we refute the suggestion that some small fraction of volcanic HCl manages to dodge the rain and commit the crime? And, while we are at it, how do we explain the Mount Pinatubo effect on the ozone hole if Suspect #3 is indeed innocent? As well as an alibi, clearing Suspect #3 will also require a special defence to explain away the influence of the volcano.

How can we distinguish positively between Suspects #2 and #3? By an argument that would have pleased Lavoisier! If Suspect #2 is guilty, then by conservation of atoms there should be other fragments from the CFCs generated together with the Cl atoms.

Collecting samples from the stratosphere is no easy task. However, as we have seen, the United States had developed very high flying aircraft back in the 1960s, at the height of the Cold War, for use as spy planes. Such aircraft were used to collect samples from the stratosphere, and it was found that *where there was stratospheric Cl, there was also HF*. There was also HCl *without* HF, but this was at a lower level than the ozone layer, supporting Suspect #3's alibi defense. Thus the HF is our smoking gun; it must have been formed by the reaction of CF fragments with water (the HF bond is the strongest single bond there is); and the Cl in the stratosphere, that is destroying the ozone, has come from the photolysis of freons.

It remains, however, to explain the Mount Pinatubo effect. The eruption was doing something. What? As it turned out, it was providing nuclei for the formation of stratospheric ice clouds, which totally alter the course of the relevant reactions.

Ordinarily, NO reacts with chlorine atoms to make the compound ClNO, thus withdrawing them from circulation, and these subsequently react with atmospheric oxygen to make chlorine nitrate, $ClNO_3$. So NO indirectly protects the ozone layer from chlorine atoms by passivating them and removing them from the catalytic cycle. Unfortunately, both chlorine nitrate and hydrogen chloride condense on the surface of ice clouds, where they react to give chlorine molecules. In the spring, when the Sun returns to the polar regions, its light splits these chlorine molecules apart to make chlorine atoms, which continue to destroy ozone, as described above.[11]

So the Mount Pinatubo effect on ozone was due to increased ice cloud formation. It had little to do with the relatively small amount of volcanic hydrogen chloride that succeeded in reaching the stratosphere. Much more important was the large amount of dust emitted, as well as sulphur compounds that produced a sulphuric acid aerosol in the stratosphere. These dust and aerosol particles provided nuclei on which ice clouds could more readily form.

Conclusion: Suspect #1, NO on its own, does destroys ozone, but *given that there is Cl up there already* the main effect of NO from jet aircraft is to scavenge Cl and thus protect ozone! Suspect #3, hydrogen chloride from volcanoes, is completely innocent, despite politically motivated efforts to frame it. Dust and sulphuric acid fumes from volcanic eruptions are, however, guilty of aiding and abetting. Suspect #2, Cl from CFCs, is guilty and the stratospheric HF is the smoking gun that proves it.

Why is the problem worst in spring? Because warming the ice clouds mobilises the chlorine nitrate and hydrogen chloride on the ice particle surfaces, and the chlorine molecules then liberated are split into atoms by sunlight.

In short, Suspect #2 supplies the chlorine atoms that do the damage; Suspect #1 partly protects the ozone from their effects; and Suspect #3 increases the abundance of ice clouds, which interfere with Suspect #1's protective action. The fluorine in the stratosphere identifies Suspect #2 as the source of the chlorine atoms, as plainly as fingerprint ever identified burglar.

Political Pressures. Whistle-Blowers Vindicated

If the ozone hole is due to human activity, we are going to have to change our ways. This might be uncomfortable or expensive. As with every environmental concern, it is easier in the short run to belittle the dangers, deny the evidence, and exaggerate the costs of change. Thus the arguments against the regulation of CFCs were very similar to the arguments formerly used against restricting whale hunting, or against the banning of leaded gasoline, and now being used against attempts to limit carbon dioxide emissions.

The technique used to exaggerate the costs is very simple. One just assumes that people will not change their behaviour, and that no substitute can be found for the offending item. For example, we were told that whale oil was uniquely suitable for some lubrication tasks, so that without it, industry would quite literally grind to a halt. We were also told that taking the lead out of gasoline would force up refining costs and reduce engine performance. In the event, substitutes were developed for whale oil lubricants, while cars have improved, and the refining process has got cheaper, since gasoline was de-leaded.

This experience has not prevented a long rearguard action against the regulation of CFCs. To this day Dana Rohrabacher, a senior and influential member of the US House of Representatives, continues to pour scorn on the scientific case for action on ozone, in order to bolster his extraordinary claims that man-made global warming (see Chapter 15) is some kind of a scam. As late as the fall of 1995, eight years after a Republican president, George Bush Sr., had committed the United States to the Montreal Protocol on the elimination of CFCs, Rohrabacher was still using his then position as Chair of the House Committee on Scientific Affairs to conduct hearings in defense of CFCs. His choice of witnesses was so bizarrely biased that it drew a rebuke from *Chemical and Engineering News*, the house organ of the American Chemical Society, in its editorial of October 9, 1995. Two days later, Nobel prizes were announced for the scientists whose work had uncovered the role of nitrogen oxides and CFCs in ozone depletion.

Why do we have ozone alert days[12] if ozone is such a good thing? How does ozone get formed at ground level anyway? Ozone *at ground level* is involved in the formation of irritating compounds. The cycle starts with the formation of NO by motor vehicles, followed by the reaction of NO with oxygen molecules to make NO_2, much as in the first steps of the Birkeland–Eyde process for making nitric acid, as described in the previous chapter. This brown gas is split by light to give NO and oxygen atoms, and these oxygen atoms will then react with molecular oxygen to give ozone, which is roughly where we came in. Notice that while in the stratosphere NO destroys ozone already there, by intercepting oxygen atoms and by removing such atoms from ozone, at the surface it promotes ozone production.[13] Ozone then reacts with hydrocarbons in vehicle exhaust to make a range of highly reactive and irritating materials, including *organic peroxides*, of type R-O-O-H, which are strong oxidising agents and irritants. Ozone can also react directly with the nasal membranes.

Passing Sentence

World production of CFCs has been phased out completely. Also found guilty of the same offence: carbon tetrachloride and 1,1,1-trichloroethane. CFCs had two main uses; cleaning flux off support boards in electronics after soldering, and as a working fluid for refrigerators and air conditioning units. New materials are being developed as refrigerants. These are HCFCs, hydrochlorofluorocarbons, which have C–H bonds that get broken by UV light long before they enter the damage zone, and HFCs, which contain no chlorine at all but are nonetheless potent greenhouse gases. Electronics companies have developed "no-wash" fluxes, which can be left in place, and can now omit one expensive step, the washing of soldered circuits with CFCs.

Is there a happy ending? Yes, it seems, eventually. For a while, it seemed that all was going well; the amounts of CFC in the stratosphere were beginning to fall. However, in April 2005 it was reported that the ozone layer was thinner than ever, although by late 2010 it had begun to recover. Why did it take so long for the ban on CFCs to have any effect? It appears that this is a result of greenhouse gases. Increasing amounts of greenhouse gases trap more heat at the surface and in the lower levels of the atmosphere, causing global warming (see Chapter 15 for the connection between stratospheric cooling and global greenhouse warming). At the same time, they block heat from reaching the stratosphere, which becomes cooler as a result. Indeed, this combination of lower level warming and upper level cooling is itself clear evidence that the warming is being driven by these gases, rather than some underlying natural cycle. A cooler stratosphere means more ice clouds, which in turn means more of the unwanted reactions that occur on the ice surface. Everything connects.

There are also many unresolved scientific issues regarding replacements for CFCs, and the role of other related substances in ozone depletion. Bromine-containing substances, in particular, are potentially even more damaging than those containing chlorine, and there is much current controversy regarding the relative roles of human and natural activity in their release into the atmosphere. Despite all this, the most recent reports[14] note definite signs of recovery in the Antarctic, with the ozone layer expected to return to its pre-1980s level by the middle of this century. However, the Arctic experienced its most serious ever recorded ozone depletion in Spring 2011, explained by an unusually long-lasting cold period in the stratoshere.[15]

Connections and Ramifications

CFCs are greenhouse gases (see Chapter 15). So are HCFCs, but these are expected to be much more short-lived because the C–H bond is comparatively easy to cleave. Ozone itself is a greenhouse gas, and ozone depletion over the Antarctic is one of the reasons why it has experienced less than its share of the general global warming of the past half-century. Once again, we see how everything connects, in ways that had not been, and perhaps could not have been, foreseen.

We live in interesting but hopeful times. The impact of growing population, limited resources, economic globalisation, and (all being well) the spread of affluence will increase the pressures of human activity on our planet throughout this century, and dealing with these will require deepening international cooperation. The agreements on CFCs are an encouraging start.

To Sum Up

The delicate thin layer of ozone in the stratosphere plays a vital role in keeping Earth habitable. Its existence depends on the balance between the processes in which it is formed, and those in which it is destroyed. In the 1980s, it became clear that the ozone layer was thinning, with potentially serious consequences. Possible reasons for this included formation of NO by highflying jets, and catalytic destruction by chlorine atoms arising from the CFCs then widely used as refrigerants. Many lines of research, including actual sampling of the stratosphere, led to the conclusion that chlorine atoms from CFCs were responsible for the damage. Paradoxically, while NO itself also causes damage, its most significant effect at present is to protect the ozone layer by removing chlorine atoms.

Things could have been far worse if the economic climate had permitted the extensive development of commercial supersonic flight.

While the American political Right blamed (and blames) the damage to the ozone layer on hydrogen chloride emitted from volcanoes, the volcano effect has a completely different explanation; volcanic dust nucleates ice clouds, which interfere with the process by which NO removes chlorine atoms.

CFCs are now banned by international treaty, providing a badly needed precedent in environmental cooperation.

8 Rain Gauge, Thermometer, Calendar, Warning

Sunlight and rainfall, the collapse of three great dynasties, and a grave warning about the future; all these inscribed in the detailed chemistry of a hands breadth height of stalagmite from a cave in China.

The Stalagmite's Story

Wanxiang Cave[1] is a limestone cavern in central China, towards the northern end of the vast and populous area that relies on the Asian monsoons for its rainfall. Warm air rises from the land surface of Asia, which has been heated by the summer sun, sucking in moist air from the Bay of Bengal and the South China Sea, which deposits its load as it rises and cools when making its way inland.

The area is semi-arid, but during the rainy season water seeps through the roof of the cave and drips down onto the floor, where it slowly evaporates. Rainwater is naturally slightly acidic, because of its carbon dioxide content, enabling it to dissolve small amounts of the calcium carbonate that makes up limestone. This is redeposited as the water evaporates and the carbon dioxide diffuses away. So, as we tell our students, a growing stalagmite serves as illustration of the principles of chemical equilibrium. What we do not tell them — perhaps we should — is that it can also serve as a rain gauge, a thermometer, and a calendar.

To use the stalagmite as a calendar, we require a method of radio-active dating. Carbon dating will not do, because the carbon in the stalag-mite will simply reflect that in the parent limestone. Potassium carbonate is soluble and washes out, so we cannot use potassium/argon. Fortunately, the Wanxiang stalagmite contains traces of uranium, and uranium/tho-rium dating (^{234}U to ^{230}Th, half-life 245,000 years) can, unlike the more familiar ^{238}U/^{206}Pb uranium/lead method, be used even when the deposit is quite recent.

The stalagmite studied in this work was 118 mm long, and had been growing continuously at much the same rate, as established by dating at various points along its central axis, since 190 AD; this gives an average growth rate of roughly 1 inch every 390 years.

To see how the stalagmite gives us a climate record, we need to look more closely at one of the simplifying assumptions that we make when teaching atomic theory. Here we tell our students that the different iso-topes of an element have slightly different physical properties, because of different mass, but identical chemical properties. This is not exactly true. Because of quantum mechanical effects,[2] there is a small but measurable tendency for the heavier isotope to be concentrated in the more tightly bound environment.

Now consider how this applies to water. The oxygen in H_2O contains about 0.2% ^{18}O, and smaller amounts of ^{17}O, in addition to ^{16}O, the domi-nant isotope, all made billions of years ago by the reaction of helium and carbon nuclei in long-dead stars. In liquid water, the oxygen atoms are more tightly held (by the hydrogen bonds between the separate molecules; see Chapter 14) than in the vapour, which as a result is relatively depleted in the heavier isotopes. The degree of depletion is dependent on tempera-ture, and the amount of ^{18}O in any sample will depend in quite complicated ways on its history. A stalagmite represents calcium carbonate deposited from a much larger amount of water, so that fluctuations in the ^{18}O content of the water are inherited by the carbonate. We can lay out the record of these fluctuations alongside other kinds of record, and see what they track.

They track the Sun. They track rainfall. They track temperature. And they track the rise and fall of empires.

Variations in solar output affect climate both directly and indirectly. The direct effect is obvious. The indirect effect operates through the Sun's magnetic field. This partly shields the Earth from the high-energy cosmic rays bombarding it from deep space. The magnetic field is strongest when the Sun itself is strongest, and also when there are active sunspots. A weak magnetic field means that more cosmic rays get through, creating ionised tracks in the atmosphere. These tracks then act as nuclei for the formation of high altitude ice clouds, which reflect more sunlight back into space. So a weak Sun, and especially a Sun relatively free of spots, will cause global cooling both directly (less energy) and indirectly (more high ice clouds). If sunspot activity does influence the isotope record, there should be a regular wobble corresponding to the known 11-year sunspot cycle. We can apply mathematical (Fourier transform) analysis to the trace, and indeed it's there.

There is other evidence for the influence of the Sun. Times of more efficient removal of ^{18}O (periods of strong depletion) turn out to correspond to periods of warmer weather throughout the northern hemisphere, as shown by independent historical records. Thus the early ninth century is marked by weaker depletion of ^{18}O in the cave record, and an advance of the glaciers in the Swiss Alps, giving way by the end of that century to strong depletion, the retreat of the Swiss glaciers, and the warm spell that made possible the Viking colonisation of Greenland. In the mid-14th century, things changed again quite dramatically, with the onset of weak depletion corresponding to the worsening weather and the abandonment of the Greenland settlements. The weakest depletion found occurred shortly after 1600, during the coldest period of the "Little Ice Age". This period has been credited for the particular excellence of Stradivarius violins, made from the denser wood that grew more slowly in cooler weather.

How does the Sun's activity affect the isotope record? Through rainfall. We can think of the evaporation of water from the tropical oceans, and subsequent rainfall on land, as an enormous distillation process, in which the vapour is preferentially depleted in the heavier ^{18}O. More sunlight on the oceans means more evaporation, heavier rainfall, more efficient removal of the heavier isotope, and greater ^{18}O depletion of the water that eventually percolates through the cave and exchanges its oxygen with the growing stalagmite.

Until very recently China was (as much of it still is) a near subsistence level agricultural economy. Such an economy is greatly affected by fluctuations in climate and especially, in the semi-arid northern part of China, in rainfall. Poor rainfall means poor harvests, and poor harvests lead to popular discontent or even civil unrest. Again the record bears this out. The periods of weakest isotopic depletion (least rainfall) occurred around 900, around 1400, and around 1600. These corresponded to the beginnings of periods of instability and discord in China's history; the "Five Dynasties, 10 Kingdoms" period of rapid dynastic turnover and national disunity from 907 to 960, the collapse of the Yuan (Mongol) Dynasty in 1368, and the transition from Ming to Qing in 1644. The second half of the 19th century was a period of strong isotopic depletion, and correspondingly good weather and strong monsoons, but this was not enough to save the last imperial dynasty, the Manchu, because by then the affairs of China had become linked to the outside world, the imperial ambitions of the European powers, and the rising power of Japan.

The Wanxiang cave study has aroused much interest (not only in China, but elsewhere), and been cited in the scientific literature over 100 times in the three years since it appeared. One interesting study[3] relates the data from Wanxiang and other caves to the growth and shrinkage of deserts in Mongolia and northern China. Times of increasing desertification matched the decline of dynasties, while strong new dynasties arose during periods of reversal. Another more recent analysis[4] compares the temperature records in northern China as a whole and in Beijing with records of other kinds of event. Colder weather throughout this period led to drought, increases in the price of rice, more frequent locust plagues (shrinking lakes create beaches on which locusts lay their eggs), internal disorder, and invasion from the North, as drought affected the livestock-dependent peoples of Mongolia and Siberia.

What of the present Chinese regime, and present era of global warming? Will this warming strengthen the monsoon, bringing more rain to China and India? Unfortunately no. From 1950 on, we have good weather records for this region of China, and we can compare these directly with the isotopic record. This still tracks rainfall as before, but the relationship with temperature has gone into reverse. For reasons not understood, what had previously been a positive correlation has turned into a negative one, indicating disruption of the previously existing feedback loops. In plainer English, as our emissions have made the planet warmer, the monsoon has weakened instead of strengthening. We don't know what it is about what we're doing, but it is disrupting the rains on which over a third of humankind depend. This at the same time as we are shrinking the Himalayan and Tibetan glaciers, which feed the great rivers that water the same area.

The Tree Ring's Tale,[5] and Other Stories

A stalagmite has got one very special advantage over other techniques. We know exactly what location it refers to. However, there is an obvious short-age of conveniently located limestone caves containing suitable stalagmites. The radioactive dating can never be precise, because of statistical uncertain-ties, and because of possible variations in growth rate in between the points at which we take our samples. But when we want to relate climate changes to events in history, we would really like to date changes to the exact year. There is, in fact, a way to do this, and that is to look at tree rings.

Every year, trees grow a ring of lighter coloured wood during the summer, and a thinner, darker ring during the winter. The pattern of tree rings depends on the weather from year to year, so much so that the tree ring pattern on ancient wooden objects can be used to date them. Tree rings are even used to calibrate the radiocarbon dates used in archaeol-ogy. Such calibration is necessary because variations in the amount of cosmic ray bombardment reaching the Earth give rise to fluctuations in the rate at which the radioactive isotope of carbon is generated.

The more favourable conditions are for a tree, the thicker the ring that grows each year. There are two variables that affect the rate of growth, namely temperature and rainfall. So extracting a climate record from tree rings requires us to disentangle these two different factors. This is done by choosing species of tree that grow under different conditions. Mountain conifers generally get enough water, so their growth rate is controlled by temperature. Trees at the fringes of hot deserts generally have enough warmth, so their growth is controlled by the availability of water. We have good records of both temperature and rainfall in Europe and North America going back more than a century, so we can use these observa-tions to validate the tree ring record.

There are numerous other methods for studying historical and ancient climates. Ice cores from the Arctic and Antarctic show annual bands, which trap bubbles of the Earth's atmosphere as it was when they were compacted. So we have combined together exact dates, information about atmospheric changes (such as historical variations in carbon dioxide content), and temperature information from the ^{18}O content of the ice. The growth and retreat of glaciers is another obvious indicator of temperature, although, like tree ring thickness, it depends on the amount of local precipitation as well as on the temperature. Sediments in lakes and on continental shelves provide soil samples from the surrounding areas, and copious run-off leads to high levels of those elements (titanium and iron) abundant in terrestrial rocks. Going back further, more indirect climate markers are used, such as types of pollen, other evidence of species habitat, chemical tracers resulting from biological activity, and overall changes in the types of sediments being laid down.

Elsewhere on the Planet

There is no shortage of other examples of climate change leading to political change.[6] A sharp and severe drought, evident from wind-blown dust deposited in the Gulf of Oman, may have brought down the short lived Akkadian Empire, precursor to the empires of Assyria and Babylon.[7] Stalagmite evidence, similar to that from China, and confirmed by analysis of Caribbean sediment cores,[8] shows a series of severe droughts on the Yucatán Peninsula during the period of collapse of the Maya civilisation, for no other apparent reason and at the height of its accomplishments.[9] The collapse of the "Lost Colony" of Roanoke, Virginia, which completely vanished between the arrival of the first settlers in 1587, and the arrival of resupply ships four years later, was at one time blamed on poor planning and inadequate supplies, but is now thought to be the result of the most severe drought to hit what is now the south-western United States in the past 800 years. Intermittent droughts in the upper Rio Grande Valley region disrupted the late 13th century culture of the Anasazi, ancestors of the present-day Pueblo people. Tree ring data from Southeast Asia bear witness to a combination of overall drought, and increased local climate variability, in the late 14th an early 15th century. These no doubt contributed to the decline of the once powerful kingdom centred on Angkor, in present-day Cambodia, where displaced masonry blocks and sand deposition in the elaborate canal system give evidence of a combination of drought, and periodic severe flooding.

Depending on species of tree and location, tree ring data can be influenced both by the amount of rain, and by temperature. An unusually painstaking study of climate change in Central Europe (8,800 separate samples examined) combined tree ring data from oaks, which reflect water availability in the spring growing period, with similar data from conifers growing at elevation in the Austrian Alps, for which summer temperature is the limiting factor. The results provide evidence of declining rainfall, and greater climate variability, in the period between 250 and 600 AD, corresponding to the period of the major migrations that led to the collapse of the Roman Empire.[10] A high resolution examination of a stalagmite from a cave in the hills south west of Jerusalem[11] involved the study of individual annual bands, detectable under the microscope by fluorescence of organic material included within the growing stalagmite, and comparison of winter and summer ^{18}O depletion levels. This revealed a continual decrease in rainfall across the period from 2,000 to 900 years before present, which the authors relate to the decline of Rome, and to the vulnerability of the Byzantine Empire to the Muslim onslaught.

These last two proposals strike me as more suggestive than conclusive, especially given the natural human tendency to impose patterns on events. Human societies are certainly vulnerable to the effects of climate, sometimes tragically so, as news programmes remind us. As I write, the Horn of Africa is devastated by drought, and such events are expected to become more common (see Chapter 15) as the climate continues to warm. However, it is not all that easy to pinpoint changes, or to separate the effects of trends from those of variability, and tracing causal connections in history is the most difficult task of all. Hence the need[12] for a combined approach, bringing together the traditional methods of history and archaeology with the new data emerging from the kind of investigations described here.

To Sum Up

In the words of one of the papers quoted here,[10] "Historical circumstances may challenge recent political and fiscal reluctance to mitigate projected climate change." Rulers of the world, beware!

9   **Making Metal**

The word "metal" comes to us from the Greek metallon, mine pit. This in turn is related to metallao, "I enquire", a word perhaps ultimately derived from meta alla. Meta as in metaphysics, alla as in allergy or allotrope or allele, other; in search of other things.[1] *Let us dig more deeply into the subject.*

The Gold Standard

"In truth, the gold standard is already a barbarous relic."

John Maynard Keynes, Monetary Reform, 1924

Why is gold beautiful? Why is it rare? Why is it sought-after? Why is it called a "noble" metal? And why was it one of the first metals to be worked, in every civilisation?

Aristotle thought that metals matured within the Earth, so that base metals would progressively become more noble. The influence of these ideas can still be seen in the language of modern chemistry. Ordinary metals, or rather their oxides, are *bases*, and so, in modern usage, are any substances that can neutralise acids, while unreactive metals are still called *noble*. Gold, in both ancient and modern senses of the word, is the noblest metal of them all.

Gold is beautiful because of its colour, and even more importantly because of the perfect shininess of its surface. The colour arises from the way in which light of different wavelengths interacts with the electrons that hold this and, indeed, any metal together.[2] The perfect shininess arises from the fact that gold is extremely unreactive, so it does not build up a film of oxide on the surface. There are reasons for this *inertness*. As we shall see in later chapters, chemical bonding depends on sharing of electrons. The particular arrangement of electrons in gold makes it easy to share them with another metal, but difficult to share them with anything else. So gold forms many *alloys* with other metals, including the alloys used in jewellery, but otherwise prefers to bond to itself, rather than, say, to oxygen or sulphur. It is the only metal that will not react directly with sulphur, and while a gold oxide is known, it can only be prepared indirectly and decomposes on heating.

There are two reasons why gold is rare. One is that it has a large number of particles in its nucleus; 79 protons and 118 neutrons, to be precise. In this particular case, we can be very precise, because gold is an element that has only one naturally occurring isotope. As a result, its atomic weight is extremely close to a whole number; 196.9665 on the atomic mass scale. As we shall see in the final chapter, all elements with atomic weights greater than about 60 are generated in red giant stars or supernovae, by the addition of neutrons to lighter elements, followed by radioactive decay. As you might expect, heavier elements are generally rarer, and there is an overall reduction in abundance with increasing atomic weight from iron through to uranium. In fact, if we look at the Solar System as a whole, the abundance of gold is just one ten millionth of the abundance of silicon.

But if we look at the abundance in the Earth's crust, gold is even rarer. Along with platinum, and a few other elements (ruthenium, rhodium, palladium, rhenium, osmium, and iridium), gold is a hundredfold less abundant in the Earth's continental crust than it is in the Solar System as a whole. It has been removed by a very efficient process.

Remember that gold can bind to other metals, and these other metals include iron. Early in the history of the Earth, a huge amount of molten iron separated out and sank into the core, taking gold, platinum, and other *siderophile*, or iron-loving, elements with it, while the *lithophile* elements, those that are readily incorporated in rocks alongside oxygen, silicon, and aluminium, remained on the surface. The siderophiles were replenished to some extent by the arrival of Earth's "late veneer", the final stages of planet accretion after the core had formed, but this only accounts for a small fraction of all terrestrial material. Gold is a siderophile because it will interact with iron but not with oxygen, and it is beautiful because of its lack of reactivity towards oxygen, so it is rare for the same reason that it is beautiful.

As to why it is sought after, that should now be obvious. Firstly, because of its beauty, combined with the fact that it never rusts or tarnishes. Secondly, because it is rare, and its rarity makes it a seemingly secure source of value. Actually, there is something rather strange about that last statement. People want it because it is valuable, but it is valued because it is wanted. For centuries, people imagined that money was valuable because it could be turned into gold. This was the "gold standard" that Keynes was denouncing in the quotation at the head of the chapter, and it did enormous harm. The amount of money required to maintain demand in an economy depends on the value of the goods and services produced, and not on such an arbitrary standard as the amount of a particular metal that happens to have been mined by that date. The obstinacy of American adherence to the gold standard as late as 1933 is thought to have helped prolong the Great Depression, and the last notional link between gold and money was not cut until 1971, under the Nixon administration. Traditionally, people flee to gold for security in times of financial crisis, but I would advise against this. In 1970, international crises sent the price of gold up to a record value, which it did not reach again for another 38 years, and in real (inflation-adjusted) terms, its present value is still only half of what it reached then. Gold pays no interest or dividends. It is used in electronics, jewellery, and decorative items, but demand for these purposes is a fraction of annual production, and the rest is only worth what someone will give you for it. The world is full of gold mines that are only marginally profitable, some working, some idle. So if it were thought likely that today's high prices were here to stay, more of these would restart production, and send the price down again.[3] To quote Steve Romick, of FPA Mutual Funds, "Gold is not an investment. A hedge? Yes, but not an investment."

Gold occurs in Nature as the pure metal, sometimes in the form of nuggets that can be collected and worked with no further processing. This is what has made it available for use since very ancient times. In a freshly discovered goldfield, little nuggets lie on the surface just waiting to be picked up. This explains the great gold rushes of California (1849) and the Klondike in the Canadian North West (1897). In the 1960s, Alaska[4] and California both adopted gold as their official state mineral. Most of the gold is dispersed as tiny flakes in rock. Some of this, as the result of erosion, may end up in alluvial "placer" deposits from which it can be separated by careful sieving or panning, with the prospector hoping for something more than a mere "flash" in the pan. After the initial rush, long-term commercial mining depends on separating the metal from the surrounding rock, by grinding and concentrating. Traditionally, this was done by treating the enriched ore with mercury, in which gold dissolves (remember how readily it interacts with other metals). Present-day treatment involves extraction with cyanide solution. The cyanide group forms particularly strong bonds to gold, so that under the combined influence of air and cyanide, gold is converted into a soluble complex salt.[5]

You may recall that Fritz Haber had plans to pay Germany's war reparations after World War I by extracting gold from seawater, but that the plan failed because his analyses showed that the concentration of the metal was too low, much smaller than had been realised. We now know that even Haber's estimate was too generous by something like a factor of 1000, and that if we could extract all the gold in the oceans (an obviously impossible task), it would add up to no more than about 10% of what we have already.

From Stone to Bronze, From Bronze to Iron

Among the elements, copper is more than 10,000 times as abundant as gold, but is still far from common, standing around 26th in order of upper crustal abundance. Yet it was the first metal to be extracted by chemical means from its ore, and in current annual production the only metals that exceed it are aluminium and iron.

Copper is chemically quite unreactive, and is sometimes found as the *native* metal. More commonly, it occurs, on its own or alongside other metals, in ores in which it is combined with oxygen or sulphur. The native metal is quite soft, useful for decoration but not much else. Hammering and rolling harden it ("work hardening"), by introducing irregularities in the crystal structure so that rows of atoms can no longer move freely past each other. If worked too much, the metal becomes brittle, but it can be made soft and workable again by heating ("annealing") All of this was known in Neolithic times, even before the invention of pottery and the domestication of animals, and the development of metalworking was one of the changes that led to the beginnings of urbanisation.

Since copper is so unreactive, its oxide is readily *reduced* with carbon, available in the form of charcoal from very ancient times. Here is how it works. To start with, charcoal (from wood), mixed with the ore, burns in a restricted supply of air. This gets things going by making carbon monoxide (CO). The carbon monoxide can then remove oxygen from the metal oxide ore, to give carbon dioxide (CO_2) and metal. If the temperature is high enough and there is enough charcoal, this carbon dioxide will react with the charcoal, to form more carbon monoxide, which reacts with more metal oxide, and so on until we run out of metal oxide or charcoal. This kind of reaction can be used to prepare metals, provided the carbon-oxygen bond is favoured over the bond between oxygen and metal.[6] If the temperature is high enough, the metal will melt and run to the bottom of the firing pit. Otherwise, it will need to be separated from ash and the slag formed from unreduced metal oxides and other impurities in the ore.

Copper sulphide can be converted to copper metal by heating it in a restricted amount of air, which removes the sulphur as sulphur dioxide, leaving the metal behind. The sulphur dioxide is then reacted with oxygen and water to give sulphuric acid, a valuable byproduct.[7] However, escaping sulphur dioxide gives rise to sulphuric acid haze in the atmosphere, and contributes to acid rain. An alternative method is treatment of the sulphide with bacteria (*biohydrometallurgy*). These convert the insoluble copper sulphide into soluble copper sulphate, and the copper sulphate can be reduced to elemental copper by treatment with a more reactive metal. Scrap iron is ideal for this purpose.[8] This methodology has been used at the Rio Tinto mine in Spain for over a century. It is particularly useful for low-grade ores, since the bacteria do the work of separating the material of interest from the surrounding rock, and will presumably become more widely used for copper and other scarce metals as high-grade ores are depleted.[9]

Copper smelting, and the more complicated iron smelting technology that arose later, are not the results of single individual inventions, but of accumulated knowledge and discovery. Neolithic man (or perhaps woman) had found that heating flint made it easier to work with, had discovered the decorative potential of malachite beads, and had developed techniques for shaping and drilling them. Native copper was also being used in Neolithic times, and frequently occurs alongside decoratively coloured ores. At some stage someone would have discovered that the copper was more easily hammered into shape when hot. We can tell if this has happened using *metallography*, etching of the metal surface to make grain boundaries and other irregularities more visible, since such treatment alters the way in which the crystals that make up the solid interlock with each other. We can also use metallography to detect the *defects* produced by cold working, and the effects of subsequent annealing, and we know that annealing and cold working were being used on native copper by the end of the ninth millennium BC in eastern Turkey.[10]

Until recently, it was assumed that copper smelting (and a great deal else) first arose in the Middle East, but increasingly this assumption looks like the product of our own cultural bias. It now seems as if copper and bronze technology were considerably more widespread in prehistoric times, and may have been developed several times independently. Balkan copper smelting, as we shall see, began 500 years earlier than anything so far discovered in the Middle East, and the volume of Balkan copper artefacts discovered also greatly exceeds that from the Middle East.

"Ötzi the Iceman", who met a violent death around 3300 BC near what is now the boundary between Austria and Italy, was carrying a copper axe, and had traces of copper and arsenic in his hair. The arsenic is significant because, as we shall see, it is suggestive of copper smelting rather than simply the use of native metal. Another famous metalworker is the "Amesbury archer", buried around 2300 BC near Stonehenge, whose grave contained a set of metalworking tools and an anvil stone, as well as sumptuous burial gifts.

Almost 2000 years earlier, copper was already being smelted in the Balkans. One site in particular, at Belovode in what is now Serbia, was in use for about 700 years, starting around 5000 BC.[11] There is on-site evidence for the use of malachite (a copper carbonate) for beads, and of different copper containing materials as ores for production of the metal by smelting with charcoal. This requires temperatures of over 1000°C and it seems that this was achieved in firing pits. The slag from these pits was glassy, showing that high temperatures had been reached. It also contained other metals as their oxides, showing that the slag was formed as a byproduct from smelting, and not just adventitiously from casting native copper. Isotopic analysis of lead impurities[12] showed that ores had been brought from various locations, and over trade routes extending for hundreds of miles.

Our word "copper" comes from the Latin *cuprum*, which is derived from the name of the island Cyprus,[13] although some philologists think that it was originally the other way round, and that the name of the island derives from the Sumerian word for copper or bronze. The large copper deposits on Cyprus also contain arsenic, so that *smelting* the ore automatically produced an alloy, arsenical bronze. In much the same way, smelting copper ores containing zinc produces brass. It is thought that the earliest alloys were made accidentally by smelting such mixed ores. However, it was realised early on that deliberately mixing two different kinds of metal ore would give rise to alloys, and that these alloys were harder than the pure metals.

As to why alloys should commonly be harder than pure metals, think about what happens when a piece of metal is bent out of shape. Things not made of metal would simply break, because the chemical bonds that hold them together depend strongly on the angles between the bonds (more on this in Chapter 12). The bonding in metals is much less dependent on direction, so that layers of metal atoms can be forced to slide over each other, and still remain bonded. That is why you can hammer out a block of metal, but you cannot hammer out a block of table salt or a lump of charcoal.

If you have an alloy, rather than a pure metal, some of the atoms will be of different type, with different size, and different local bonding. As a result, they will not slide past the other atoms so smoothly, so you will need to exert more force to deform the metal, and the metal will be tougher and harder.

Pure copper is too soft to make knives, spear points, or helmets, but bronze will do very nicely. There are, however, a couple of problems with arsenical bronze. It only works for one particular kind of copper ore, and the metal smiths using it don't live very long. That's serious. Smiths, as we see below, were a valuable military resource.

Most bronze was made by alloying copper with tin. Tin oxide ore originates in granite, and accumulates in soils and gravels produced by granite weathering, where it is easily detected by its dark colour. Like copper oxide, it is easily smelted, so that copper-tin bronze is easily prepared from the mixed ores, or by melting the metals together. Not only is this bronze much harder than copper, but it has a lower melting point, making it easier to separate and cast. The trouble with tin, however, is that it is considerably scarcer than copper.

The sources of tin in the ancient world are still uncertain.[14] Some tin bronzes will have occurred spontaneously, through the occurrence of tin in association with copper ore, but this is exceptional. Small local deposits were exploited, but there was also long-distance trade, involving tin mines in Central Asia, on what later became known as the Silk Road through present-day Uzbekistan, and also mines in what is now the border region between Germany and the Czech Republic. Isotopic and trace metal analysis show that deposits in what is now western Iran supplied ore to Iran and Mesopotamia.[15] The world's oldest known shipwreck (found near the Turkish coast, and dated to around 1400 BC) contained both copper and tin ingots in its cargo, showing that by this stage the two metals were being deliberately blended together. The Romans, and perhaps the Phoenicians before them, exploited deposits in north-western Spain, Brittany, and Cornwall. These last deposits derive from the massive granites of Dartmoor. The Cornish mines were a major resource from ancient times until relatively recently, and included multilevel mines, and galleries extending beneath the sea floor. Only in the 19th century was Cornwall superseded by newly found ores in Bolivia, and in a broad arc in East Asia stretching from China to Malaya.

China is now the world's largest supplier of tin, and home to more than half the known economically usable reserves. Most of the world's copper production is from Chile, which also has the largest reserves of that metal, closely followed by Peru. Copper is extensively used, because of its corrosion resistance, excellent electrical and thermal conductivity, and ease of working, for electrical wiring and for piping. While tin-copper bronze has mainly been replaced by brass and ferrous metals, tin is still used for protective plating and the making of alloys, such as solders. Liquid tin also supplies the basis for float glass manufacture, which has been the standard method of plate glass and window glass production for many years. Unless major new deposits are found, which seems unlikely, tin can be expected to run out in a few decades at current rates of usage, and copper by the end of this century.

Something like 4% or more of world tin production comes from areas under the control of renegade militias, in the unhappy country that calls itself the Democratic Republic of the Congo.[16] At the time of writing, the US and European governments are considering ways of controlling trade in the "three Ts", tin, tungsten, and tantalum, from the Congo conflict zones. This suggestion is attracting widespread political support from both Left and Right. No one wants to be subsidising gang rape when they buy a cell phone.

Corinthian bronze, which shone like gold, was used for the gates of Herod's Temple in Jerusalem, and seems to have been a gold-containing copper alloy, treated to remove copper from the surface, leaving a layer of gold behind. This bronze was known in North Africa, suggesting a historical connection with transmutation into gold, and the rise of alchemy.[17]

Compared with copper and tin, iron is extremely common. It is among the most abundant elements, both in the Solar System as a whole and in the Earth's crust. Its oxides are responsible for the red-brown-yellow colours of rocks and soils, and occur in massive thick deposits, known as banded iron-formations, in many parts of the world. In terms of nuclear binding energy, it is the most stable element, and the ultimate product of nuclear fusion in stars (see final chapter). However, the "Iron Age" did not begin until more than 2000 years after the "Bronze Age". This is because iron is much more difficult to smelt, and its properties much more difficult to control.

Despite these difficulties, which we describe in more detail below, small amounts of iron have been used all over the world for almost as long as any other metal, and Captain John Ross, looking for the North West Passage in 1818, was amazed to discover that the Inuit had iron tools.

How is this possible? It is possible because the metal was, literally, out of this world. The Sumerian word for iron was *anbar*, combining the two roots *an* (sky) and *bar* (metal), and the metal they were familiar with from the earliest times was meteoritic.[18] Meteorites (with a few interesting exceptions) are fragments that have broken off from asteroids, and end up getting captured by the Earth's gravitational field. Most of these are stony, but about 5% are made of iron, fragments of the cores of asteroids that were molten early in the history of the Solar System. Meteoritic iron contains nickel, as shown by Smithson Tennant in 1806 and again by William Wollaston (yes, him again! See Chapters 4 and 5) in 1816. In fact it is fairly easy to distinguish from terrestrial iron, because the nickel content is upwards of 5%. This incidentally makes it, like alloys in general, harder than the pure metal, and therefore more useful for toolmaking. Meteoritic iron was rare, and much more valuable than silver.

The syllable *bar* lives on in words for brass in modern European languages, and, via the Latin form *ferrum*, in the Romance language words for iron, as well as in English trade and technical terms (farrier, ferrite). It was elaborated into the ancient and modern Hebrew *barzel* for iron, and related words in other Semitic languages, which possibly (via the Hebrew/Phoenician and the known metal trade routes) gave rise to the unusual English words brazil for iron pyrites, and brazil wood for a tropical hardwood.

There are many reasons why iron is more difficult than bronze to work with. Iron is higher melting than copper (1535°C as opposed to 1083°C), and the melting point of copper is greatly lowered if the ore mix used contains other metals. Even without melting, the conversion of iron ore to metal requires higher temperatures, which could only be obtained using bellows. This gave the metal in the form of a mass or "bloom", requiring repeated hammering to separate it from slag. The pure solid iron made in this way is softer than bronze. It is easy to beat into shape, but will not keep a sharp edge. However, if the bloom is left in contact with the carbon, some of the carbon slowly diffuses into the hot iron, making it much harder. The material eventually obtained depends on how much carbon has been incorporated, and exactly how the carbon atoms are trapped inside the metal. This in turn depends on heat treatment, and whether any of the dissolved carbon has been burnt off in air. The most carbon-rich material (pig iron) forms if the iron actually melts in contact with excess carbon, since it then reacts with it chemically to make a material, known as cementite, that is hard but brittle.[19] Keeping cementite above 600°C leads to further changes, with carbon separating out as graphite. Iron containing small amounts of carbon (steel) is intermediate in properties, while iron low in carbon (wrought iron) is easily hammered into shape, but soft. Yet another important crystalline form of carbon-containing iron, martensite, is obtained by heating above 723°C, and subsequent *quenching*, or trapping the material in this high temperature form by rapid cooling. This is hard and will retain a cutting-edge, but the products of slow cooling are more resilient to shock, and preferable for the body of a tool or weapon. For the best combination of properties, quenching must be accompanied by tempering, a process of reheating and more gradual cooling. As if this were not enough, the effects of impurities in the ores are unpredictable, and traces of arsenic, which transform copper into useful bronze, make iron more brittle and difficult to work with.

If this sounds complicated, that's because it is complicated. The skill of the ironsmith, from the earliest days to the present, has consisted in adjusting the amounts and distributions of these subtly different compounds and *composites*, and key innovations seem to have arisen independently in Europe, India, Central Asia, and (earliest of all) China.[20] The preparation of Damascus or Toledo steel, or samurai swords, involved elaborate rituals of hammering, bending, tempering and quenching, and exposure to charcoal. This produced a microcomposite material in which the hard carbon-rich components, that would retain a sharp cutting edge, were united with more flexible low-carbon steel that could absorb shock without shattering.[21]

Iron production from the ore is commonly said to have originated in Anatolia in the second millennium BC, and to have spread out from there, with the Hittite Empire using iron for weaponry and trade.[22, 23] However, there are reports[24] of an earlier, and presumably independent, iron-smelting industry in the plane of the Ganges. Evidence of an early independent development in West Africa seems questionable, although the smiths of that region were certainly highly skilled and innovative.[25]

In fact, it is not possible to say exactly when and where each of the crucial developments took place in the lead-up to iron displacing bronze. Iron artefacts from the earliest period are understandably rare, since the metal corrodes, thereby destroying the evidence. There are also general difficulties in dating exposed sites, and in ensuring that radiocarbon-dated objects are truly contemporaneous with the metal finds. Nonetheless, there are numerous references to iron in Hittite and Sumerian texts, and each of these languages had more than one word for ferrous metal — as do we ("iron" and "steel") — suggesting the products of different degrees of refinement or working. In any case, it seems clear that iron was becoming more plentiful and cheaper by around 1200 BC, and replacing bronze for most purposes by around 1000 BC, throughout the Middle East and beyond. Before this change, iron is worth six times its weight in gold. Afterwards, iron is weighed by the *mina*, roughly equal to a pound, while gold is still weighed by the shekel, which is about 1/4 ounce (the shekel, like the pound, was a measure of weight before it became a unit of currency).

In Homer's Iliad, probably composed early in the Iron Age but describing events shortly preceding it, the swords are made of bronze, but there are references to iron cauldrons as princely gifts, comparable in value to horses or women. Iron axes are the prizes in an archery contest, and iron is used as a store of moveable wealth, for the axles of bronze-wheeled chariots, and in agriculture. The Odyssey, in one gruesome passage, compares the blinding of the giant Cyclops Polyphemus with a burning wooden stake to the quenching of red-hot iron.

We can also see evidence of the transition from Bronze Age to Iron Age in the Old Testament account of the conflict between the Israelites and Philistines. We know from their pottery that the Philistines had strong cultural connections with the Myceneans of late Bronze Age and early Iron Age Greece and Cyprus. We are shown their champion Goliath with a bronze helmet, but an iron-pointed spear, and told (not necessarily reliably; the authors of these ancient texts had their own political agendas)[26] that the Philistines maintained a technological edge by deporting all the smiths from Israelite-controlled territory. But by the time we come to Deuteronomy, written several centuries after this conflict, the Israelites themselves[27] have already been an Iron Age people for some time. The mineral resources of the Promised Land included iron as well as copper, and iron tools are presumably commonplace, since it is necessary to forbid their use in constructing an altar. Excavations at Taanach,[23] in what is now the West Bank, show that the site was used for both copper- and iron-smelting by the 10th century BC, and close examination of implements made there provides evidence of sophisticated procedures of alternating carburising and forging, creating items technically superior to known Philistine artefacts. Exactly when Taanach became Israelite territory is an interesting question, but what is clear is that iron working techniques were spreading and the knowledge diffusing in several directions.

Innovation in iron working has continued throughout the whole of history and into our own time. One essential change was the switch from charcoal to coke as the source of carbon, when the demand for fresh wood could no longer be met. This happened almost 1000 years ago in China, but not until the 17th century in Europe. The Railway Age demanded huge amounts of low-cost, low-carbon steel. This was made possible by the Bessemer process, which involved the addition of controlled amounts of air or pure oxygen to the molten metal, removing much of the carbon as carbon monoxide. Modern steel production techniques rely on the same basic principle. The technology continues to progress, with more than a dozen journals devoted to various aspects of metal production, and over 150 US patents granted in 2009 that mentioned "steel" in their titles.

The Poisonous Metals; Arsenic, Mercury, Lead

We have already met arsenic as an essential but accidental component of early bronze. It commonly occurs combined with sulphur, and sometimes another metal. Heating these in air converts the sulphur to sulphur dioxide and the arsenic to its own oxide, which is easily reduced by charcoal or other sources of carbon. The main source is the mineral arsenopyrite, FeAsS. This when heated decomposes to give the iron sulphide FeS, while the arsenic is given off as a vapour, which condenses in the chimney of the reaction vessel. The alchemists knew that blending arsenic with copper gave a white alloy, resembling silver, a noble metal and perhaps, in their minds, already on the way to being converted to gold.

Arsenic[28] forms alloys with many metals, and small amounts are added to brass plumbers' fittings to increase their chemical inertness, and to harden the lead plates in car batteries. It was formerly used in solder, leading to poisoning by moonshine which leached it out of joints in the still. The compound gallium arsenide is a semiconductor with uses in electronics, and the more common silicon-based electronics rely on the controlled addition of arsenic to generate electron-rich conducting pathways.

White arsenic, As_2O_3, was used in 18th century wig powder (where it must have been very effective against embarrassing personal vermin) and in the next century Victorian women used it in makeup, and even ate it (in small amounts!) to produce a pallid complexion. It was well known as a poison, useful for ants, rats and mice, unwanted spouses, and even royalty. Nero is said to have used it to get rid of his more popular younger brother Britannicus, and Cardinal Ferdinando de' Medici almost certainly[29] used arsenic to get rid of his own inconvenient brother, Francesco, Duke of Tuscany. The much-married Lucrezia Borgia (daughter of Pope Alexander VI) was also accused of arsenic poisoning, among other lurid crimes, but this information comes from politically biased sources and may just be Renaissance tabloid gossip. Be that as it may, family use declined after 1840, when the newly developed Marsh test[30] was successfully used to prove that Mme LaFarge had indeed poisoned her husband. Suggestions that Napoleon was deliberately poisoned with arsenic do not stand up to examination. He will have been exposed to arsenic from the Paris Green (copper arsenoacetate) pigment then widely used for wallpaper, but the large and highly variable amounts present in hair samples seem to have come from the subsequent use of arsenic-containing preservatives.[31] Chronic arsenic poisoning by contaminated groundwater is, however, a serious problem in many parts of the world, in particular Bangladesh.[32]

One group of arsenic compounds, arsenates, shows a chemistry very similar to that of phosphates, and phosphates play an essential role in life, not least in providing the backbone linkages of DNA and RNA. Arsenic is poisonous because it gets caught up in the same metabolic cycles as phosphorus, but the arsenic-containing molecules react with water and fall apart before they can do their job. It was therefore a great surprise when NASA announced in December 2010 the discovery of a strain of bacteria that could actually incorporate arsenic into their DNA in place of phosphorus, and continue to function and reproduce.[33] The claim was met with skepticism, on many grounds, even before it graduated from electronic version to print, and continues to give rise to controversy.[34]

During World War I, the Americans produced Lewisite, an organic arsenic compound, for use as a blistering agent, but it was impossible to disperse it under battlefield conditions,[35] and the material was destroyed in the 1950s. Arsenic compounds were, however, part of the cocktail of herbicides that the Americans used in the Vietnam War. As often happens, modified poisons have uses in medicine, and Salvarsan, also known as Arsphenamine, was the first modern antimicrobial drug, being used against syphilis from 1910 until it was replaced by penicillin in the 1940s. Arsenic-containing pharmaceuticals remain of potential value in treating trypanosomiasis (sleeping sickness, spread by the tsetse fly) and some acute leukaemias.

Precious metal prices have risen sharply in recent years, as nervous investors consider them a safe store of value in uncertain times. Mercury has shared in this rise, although it is not itself regarded as precious. Silver is slightly more abundant than mercury in the Earth's upper crust, but more than 15 times as expensive. That is because silver is geochemically *compatible* with many other elements, meaning that it will occur alongside them in the same mineral deposits, making it difficult and expensive to isolate. As we shall see, it is an important and valuable impurity in lead ores. Mercury, however, is *incompatible*; it does not fit in well with compounds of other elements, so it is easily isolated. Its major ore, mercuric[36] sulphide, is formed in volcanic fissures, and so where it does occur, it occurs in relatively high concentration. Mercuric sulphide is a bright red mineral, cinnabar, traditionally used as the beautiful pigment vermilion. The metal is easily obtained by heating the ore in a limited supply of air, when the sulphur is removed as sulphur dioxide, and the mercury distils off.

The price of mercury has fluctuated enormously, as both demand and supply have changed. One major mine at Almadén, in Spain, closed in 2003 after more than 2000 years of use, and the last US mine closed in 1992, as the richest veins of ore were worked out, and safety considerations made extraction less attractive. Demand for mercury was also falling at that time, as its use in the chloralkali industry (more about this later in the chapter, when we discuss sodium) diminished. However, it is still used in the small-scale extraction of gold. The recent rise in the price of gold has brought small, marginal mines back into production, at least for the time being, and this is what has driven up the price of mercury to its present levels. Small amounts of mercury are also used in fluorescent lamps, which are considerably more energy efficient than the incandescent lamps that they are now replacing. Production of these lamps, as of so many things, is concentrated in China, providing an incentive to develop new cinnabar mines there, with all that implies in the way of health hazards.

Mercury has the lowest melting point and boiling point of any metal. It was until recently widely used in scientific measuring and gas handling equipment, although it is increasingly being replaced because of its toxicity, and the availability of electronic sensors. The traditional unit of pressure, the torr, was defined to match the pressure exerted by 1 mm of mercury, and is named in honour of Evangelista Torricelli, who invented the mercury barometer in 1644. Why mercury? Because of its high density. Mercury is 13.6 times as dense as water, so the height of the mercury barometer is around 760 mm (near enough 30 inches), while a barometer made of water would have to be over 30 feet high. Mercury is highly poisonous, damaging the lungs and kidneys, and the long-term neurological damage sent traditional felt workers "as mad as a hatter". Pouring liquid mercury around, as I can attest, is a delightful experience. The weight, the shininess (the clean mercury surface is an almost perfect mirror), the ripples in the surface … but no safety officer could now permit it.

Mercury readily forms alloys with some metals, but not with others. One metal with which it does not form an alloy is iron, and for that reason it is conveniently stored in iron flasks (glass would burst under the weight). The standard 76 pound (34.46 kg) flask is the unit in which mercury is traded. It does, however, very readily form alloys (amalgams) with copper, silver, gold, tin, and sodium. This has made it a useful solvent for the extraction of gold and silver from their ores. The Spanish in Mexico worked this process to extract silver, no doubt at a huge cost in suffering to the indigenous workers. Dental amalgam contains copper, silver, gold and mercury. I have no qualms about carrying it in my mouth, since amalgams and other alloys are really chemical compounds, and the tendency for the amalgam to release free mercury is negligible.

Lead is another element that has been known since ancient times, and occurs in Nature as its sulphide. Roasting this in air gives lead oxide and sulphur dioxide. The lead oxide is easily reduced to the metal using charcoal or, in modern times, coke, has a low melting point (327°C), and (here I simplify slightly)[37] runs off to the bottom of the furnace. It is also possible to go directly to the metal, by roasting in a restricted supply of air. Lead prepared using these processes carries with it any silver that was present in the ore, and this may well be worth retrieving. This is done using a process invented in 1850, involving liquid–liquid extraction. Liquid zinc and liquid lead do not dissolve in each other. Silver is much more soluble in liquid zinc than it is in liquid lead. So the liquid lead is treated with liquid zinc, and the zinc in that layer is distilled off for reuse, leaving the silver behind. There was a time when benefactors were going round old churches, and generously replacing centuries-old lead roofing. I think you can see why.

Nearly all the lead now present on the Earth is the end product of radioactive decay. Lead-204 dates back to the formation of the Solar System, but the much more abundant isotopes with mass numbers 206, 207, and 208 also derive from uranium-238, uranium-235, and thorium-230 respectively. This means that different lead deposits, with different ultimate parentage, will contain different ratios of the various isotopes. This information can be used by archaeologists, when it comes to identifying where the ore for lead-containing objects came from. The technique will no doubt be more exploited in the future in tackling such vexing questions as the origin of very early tin or bronze, which frequently contain lead as an impurity.

The Romans used lead on a massive scale, not seen again until the Industrial Revolution, and we can see direct evidence for this in the amount of lead present in 2000-year-old samples from the Greenland ice cores.[38] One major use was for water pipes, and this has led to the legend that the Roman Empire fell "because" of lead poisoning in the Imperial City's water supply. This does not seem at all likely,[39] if only because the hard local water will have deposited a protective layer of limestone within the pipes. It has also been suggested that the emperors themselves, most especially Caligula, behaved in the extraordinary way recorded of them as the result of using lead salts to sweeten their wine. Again, the legend will not bear examination. We know of Caligula mainly through the reports of succeeding generations who needed to justify his overthrow, much as the popular notion of Richard III of England is based on Shakespeare, who needed to justify Henry Tudor's thoroughly illegal rebellion. In fact, the Romans were well aware of the symptoms of lead poisoning. There may, however, have been some chronic poisoning through the use of grape juice concentrate, causing occasional symptoms from Roman times onwards and emerging again as a major problem in 17th-century France. This concentrate was routinely prepared in lead vessels, and when made in this way was found to act as a preservative by inhibiting the growth of bacteria (no wonder!)[40]

In our own time, illicit goldmining in Nigeria has led to epidemic-scale lead poisoning of children, who are the most vulnerable to this cumulative poison. The rock from which the gold is mined contains appreciable amounts of lead, and weathering elutes this from the rock into the water supply. The World Health Organisation and UNICEF are investigating the scale of the problem as I write these words, and it is estimated that around 18,000 people have been affected, and that hundreds of children have died.[41]

The compound tetra-ethyl lead was for many years used as an anti-knock additive to control fuel burning inside car engines, but has been phased out because the lead from car exhaust is suspected of damaging the nervous system, especially in foetuses and young children. By a singular twist of fate, Thomas Midgley, who developed this anti-knock, was also responsible for the CFCs that, decades later, were shown to be inflicting unacceptable damage on the ozone layer (see Chapter 7).

Lead has been used in paint and until quite recently in plumbing (*plumbum* is Latin for lead), and other applications where a dense low-melting easily cast metal was desirable. The biggest use now is in car batteries, and lead consumption is rapidly increasing as car ownership spreads in developing economies. It is roughly as common as tin in the Earth's crust, and is very easily recycled. There are enough known reserves for about 40 years or so at current net rates of consumption.

Getting More Reactive; Zinc, Magnesium, Sodium

Zinc is slightly more abundant than copper, and commonly occurs along-side it in sulphide ores. Its reactivity, and the strength of the bonds that it forms with oxygen, place it at the very limit of the metals that can be prepared by smelting. As usual, the process starts by roasting the sulphide, converting it to the metal oxide and sulphur dioxide. Carbon monoxide reacts with solid zinc oxide at high enough temperatures, lying above the boiling point of zinc, to give an equilibrium mixture of gases including zinc (as vapour) and carbon dioxide. This mixture then needs to be rapidly cooled, out of contact with air, to condense out the metal. With slower cooling the reaction will simply reverse itself, giving a voluminous deposit of zinc oxide, known to the alchemists as "philosopher's wool". However, smelting zinc and copper ore together will give brass. This will initially have happened accidentally, much like the discovery of bronze, but by the end of the second millennium BC brass objects were being made with 23% zinc, which could only have been accomplished by deliberate mixing, and by Roman times this was being done on a large enough scale for brass to be used in coinage, a use revived in our own time for the Euro and UK Pound coins. (Note however that apparent biblical references to "brass", e.g. Genesis 4: 22, are uninformative because the word translated as brass is the same as the word for copper.)

Nowadays, brass is made directly by melting copper and zinc together, which means that the zinc must be prepared from its ore and rapidly condensed out, as described above. The isolation of zinc metal on a practical scale was accomplished in Asia several centuries earlier than Europe, and as late as the 17th century, Britain was importing zinc metal from India; a reminder, if such be needed, of how recent and fragile the technological supremacy of the West really is.

The biggest use for zinc is in *galvanising* steel to protect it from corrosion. Because zinc is thermodynamically more reactive than steel, the zinc will corrode first. However, the zinc rapidly builds up a protective layer of zinc oxide, so it corrodes very slowly. This is a paradox but it works.[42] The other main uses are in small cast objects such as toys, in brass-making, and in old-fashioned zinc batteries. Since 1982, US "copper" coins have been made of zinc, with a copper coating. Zinc is an essential element in diet, since it plays a vital role in many enzymes. This is related to the fact that zinc hydroxides are on the borderline between acid and base, and can react with either. Excess zinc, however, is harmful, and can interfere with the uptake of copper, which is also an essential trace element in diet. It is particularly poisonous to parrots, which may die as a result of chewing on galvanised wire mesh, or drinking acidic fruit juices from galvanised dishes.

Why "galvanised", when the usual way to prepare galvanised steel is to hot dip it into molten zinc? The original technique was to deposit the zinc electrochemically, a process named in honour of Luigi Galvani who in 1771 discovered that a frog's leg accidentally completing an electrical circuit between two unlike metals was galvanised into action.

Some zinc is still made on a small scale by the traditional distillation technique. Most is now made by *electrolysis*, which simply means using electricity, rather than carbon or carbon monoxide, as the reducing agent. The process starts by treating the ore with sulphuric acid, which reacts with the zinc compounds present to give zinc sulphate in solution. A solution of zinc sulphate does not actually contain any "zinc sulphate" molecules, but positively charged zinc and negatively charged sulphate *ions* moving independently of each other. Forcing an electric current through this solution converts the zinc ions to metallic zinc at the negative *electrode*, in the reverse of the reaction of zinc that takes place in an old-fashioned zinc-carbon battery. One unavoidable rule is that if something is reduced, something else must be oxidised, and what happens is that at the positive electrode water is oxidised to oxygen. These reactions regenerate sulphuric acid, which is then used to treat more ore.[43]

Moving to more reactive common metals, we come next to magnesium. This is an abundant element on Earth, easily formed in stars in their red giant phase (see final chapter), and occurs in many common minerals. Talc (as in talcum powder) is a magnesium silicate, with groups of atoms arranged in sheets that easily slide over each other, hence its lubricating ability. The mineral dolomite, so called in honour of the 18th century

French geologist Dieudonné Dolomieu, is a calcium magnesium carbonate, and occurs in such massive quantity that a mountain range in the Italian Alps near the border with Austria is named after it. Lavoisier (see Chapter 3) was familiar with magnesium oxide and understood its relationship to magnesium carbonate, but thought that it was an element because it would not react with carbon. The situation with calcium was very similar, so Lavoisier also regarded "lime" as an element. Sir Humphrey Davy was the first to isolate both calcium, and magnesium, by electrolysis of their molten chlorides.

Davy's electrolytic method dominated magnesium production throughout the 20th century. The process starts by isolating magnesium (as its hydroxide) from seawater, or more concentrated brines. The seawater is treated with quicklime, calcium oxide, which dissolves in water as calcium hydroxide. Magnesium is present in the seawater as magnesium chloride, which reacts with this to give magnesium hydroxide, a very insoluble material that slowly separates out as a fine powder. If you visit a traditional magnesium hydroxide preparation facility, you will see enormous shallow pools with gently sloping sides, in which this settling is taking place. The magnesium hydroxide is collected, and reacted with hydrochloric acid to make magnesium chloride. Finally, electrical current is forced through molten magnesium chloride, breaking it down into metallic magnesium at one electrode, and chlorine gas at the other.[44] (Unlike the electrochemical production of zinc, this cannot be done in water. The voltage required to reduce the magnesium ions would have to be so large that it would reduce the water to hydrogen instead.) This process was operated on a major scale by Dow Chemicals until 1998. Other companies have also worked the process, among them Norsk Hydro, the same company that was using cheap hydroelectricity to fix atmospheric nitrogen in Chapter 6.

Magnesium is so reactive that it reacts with boiling water to make magnesium oxide and hydrogen. It can "burn" in nitrogen to make magnesium nitride (hence the use Ramsay made of it in his search for minor components of air; see Chapter 5), and in carbon dioxide to give magnesium oxide and carbon. So it is not possible to make magnesium using carbon as the reducing agent, because the reaction goes the wrong way. However, magnesium can be prepared from the oxide, using silicon as the reducing agent. Oxygen bonds far more strongly to silicon than to carbon, but even so the reaction between magnesium oxide and silicon to make magnesium metal and silicon dioxide (silica) is energetically uphill. Despite this, the reaction is possible because silicon and magnesium oxide are solids at the temperature of the reaction, while magnesium, a relatively low boiling metal, can be removed as vapour. Formation of a gas is always favoured by entropy, while removing the magnesium metal under partial vacuum shifts the equilibrium in its favour, according to Le Châtelier's Principle.[45] This process is operated mainly in China, which now produces 60% of the world's magnesium, amid repeated allegations and counter-allegations regarding quota fixing and dumping.

Magnesium, alone or alloyed with aluminium, is used for lightweight construction at every level of technical sophistication from racing bicycles to racing cars to aerospace components to portable phones. It is used, as we shall see, in the production of other metals. In solution as the Mg^{2+} ion it binds strongly to phosphate groups, and plays an essential role in nucleic acid chemistry (nucleic acids are held together by a sugar phosphate backbone) and in some 300 enzymes. The magnesium ion also sits at the centre of the chlorophyll molecule, the key to photosynthesis (see Chapter 13).

Sodium is among the most reactive of all metals. It reacts violently with water to make hydrogen, generating so much heat that it melts and the hydrogen can catch fire. For safety reasons, it is stored under oil. It was first isolated by Sir Humphrey Davy in 1807, by electrolysis of molten sodium hydroxide, and is now produced by electrolysis of molten sodium chloride, common salt.[46] This is readily obtained from seawater or from "rock salt" deposits formed during the drying up of ancient seas. It is an essential component of diet, but virtually all diets contain more than enough, and too much salt in the diet can perhaps lead to fluid retention and high blood pressure.

Although sodium itself, for obvious reasons, cannot be obtained by electrolysis of sodium chloride in water, it is possible to make sodium amalgam (a dilute solution of sodium in mercury) in this way. This was for many years the basis of the chloralkali industry, which produces chlorine and sodium hydroxide (caustic soda). The sodium amalgam process is now being phased out in favour of diaphragm cells, in which hydrogen and chlorine are produced in separate compartments by electrolysis of sodium chloride in water. Removing the elements of hydrogen and chlorine leaves sodium hydroxide behind.[47]

Sodium, like all metals, is a good conductor of heat, but unlike most metals has quite a low melting point (98°C), and its alloy with potassium is liquid at room temperature. This sodium–potassium mixture is used as the primary coolant in nuclear plants, carrying away heat from the core. Heat-exchange coils containing the mixture are used to boil water under pressure, raising steam that drives the turbines that generate the electricity.

Special Tricks

As we have seen, sodium and magnesium can be prepared by passing electrical current through their molten chloride salts, and since we can make the voltage as high as we like, we can (at a cost in energy, of course) apply enough brute force to do this, despite the fact that the metals are so reactive. There are two important metals that we can't prepare in this way, for various reasons.

Aluminium occurs in many different kinds of "aluminosilicate" rock, and in the clays formed when they erode. Over long periods of time, these react with water. The silica and other components slowly dissolve away, leaving bauxite, which is impure aluminium hydroxide. We could react this hydroxide with hydrochloric acid, to make a solution of aluminium chloride in water, but oddly enough we cannot get to the metal in this way. We can crystallise out a hydrated aluminium chloride, $Al(H_2O)_6Cl_3$, but we cannot get rid of the water molecules by heating. The bond between the aluminium ions and the oxygens in the water is so strong that if we try to do this, we lose the hydrochloric acid along with some of the water, and are left with aluminium oxide. So the chloride electrolysis method, as used for sodium and magnesium, is not available.[48]

Aluminium compounds had been known, and used in dyeing, since classical times. Davy, reasoning by analogy with magnesium, realised that they must contain a hitherto undiscovered element. Aluminium chloride free from water can be prepared by passing a current of chlorine gas over a mixture of charcoal and aluminium oxide at red heat. The chlorine reacts directly with the charcoal to make carbon tetrachloride, and this then reacts with the aluminium oxide to give aluminium chloride and carbon dioxide, a thermodynamically favoured reaction.[49]

However, aluminium chloride cannot be electrolysed to give aluminium metal. It is on the borderline between the "ionic" compounds discussed in Chapter 12, which are built up of positively and negatively charged ions, and "covalent" compounds held together by electron sharing. Solid aluminium chloride is an ionic compound, but on heating it vaporises to form molecules with formula Al_2Cl_6. In these, two of the chlorines act as bridges between the aluminium atoms, so that each aluminium is bonded to four chlorines, a particularly stable arrangement as we shall see. The only way to convert aluminium chloride to aluminium metal is to treat it with an even more reactive metal, such as potassium or sodium. The result was that aluminium was not isolated until 1825, and remained an extremely expensive metal for many decades. Napoleon III reserved his aluminium cutlery for very special guests; less distinguished folk had to make do with gold.

When aluminium was used to cap the Washington Memorial, in 1884, the metal was still as expensive, gram for gram, as silver. By 1893, the price had fallen to the point where it could be used for the world's first full-size cast aluminium sculpture, the statue of Eros in honour of the Earl of Shaftesbury in London's Piccadilly Circus. A suitable subject for what used to be the hub of London's red light district, except it isn't really a statue of Eros at all. It's his twin brother, Anteros, who represents love returned, and is meant to symbolise the Earl's selfless love of the poor. Whether the poor returned this love is not known.

What happened in between these times was the invention, simultaneously in America and in France, of a way of producing aluminium by electrolysis of its oxide.[50] This oxide has a very high melting point (2072°C), but dissolves in molten cryolite, a mineral of composition K_3AlF_6, potassium aluminium fluoride. The mineral itself is rare, but synthetic cryolite is easily made.

Our final example is titanium, the "space age" metal. It is the tenth most common element in the Earth's upper crust, and occurs widely as oxide ores, often in combination with other metals. It was recognised as a separate element as early as 1791, but the pure metal was not prepared until 1910. The metal has a high melting point (above 1650°C), and the highest strength to weight ratio of any metal, while its alloys, especially with aluminium, perform even better. As a result, it has extensive use in all kinds of aircraft, and there are some 45 tonnes of titanium in a Boeing 747. Titanium dioxide can easily be prepared in the form of small, strongly light-scattering particles (titanium brilliant white), the basis of paints, and

90% of all titanium is used in this way. Other niche uses of the actual metal are in spectacle frames, and in surgical implants.

The difficulties in producing titanium metal are very similar to those for aluminium, and titanium production today is in some ways roughly where aluminium production was in the 1880s. It is impossible to prepare titanium metal by reduction of the oxide with carbon, because this reaction gives rise to titanium carbide, TiC, an extremely hard material. In fact, high-quality drill bits are sometimes deliberately coated with titanium carbide, by exposure to carbon. So the method currently used is almost the same as the method that was used originally for aluminium. Heating the oxide with carbon and chlorine gives titanium tetrachloride. This is the perfect material for making titanium white pigment, since it reacts with water to give finely powdered titanium dioxide. However, unlike the chlorides of magnesium and sodium, it cannot be electrolysed to the metal because it is made up of $TiCl_4$ molecules, and does not conduct electricity. So, just as in the old method for producing aluminium, the titanium tetrachloride is reacted with magnesium metal, to give magnesium chloride and metallic titanium.[51]

The "obvious" trick is to find a suitable solvent for titanium dioxide, much as cryolite is used as a solvent for aluminium oxide, and there are several different research groups racing to find such a solvent mixture. One ingenious variant (known as the FCC Cambridge Process) uses molten calcium chloride. Electrolysis produces calcium metal at the negative electrode, and this continuously reacts with titanium dioxide to give titanium metal and a solution of the calcium oxide in the chloride. Much as in the case of aluminium, the oxygen is released at positive electrodes made of carbon, with formation of carbon monoxide and dioxide.

Why all These Differences?

Comparing the different metals, it may have occurred to you to wonder why some are much more reactive than others. The answer varies from case to case. The reactivity of a metal will depend on two things; how strongly it bonds to itself (the more strongly, the more difficult to get it to react, and the easier it will be to reduce its compounds to the metal), and how strongly it bonds to other things (fairly obviously, the stronger these bonds are, the more difficult it will be to break them, so the metal will be more reactive and its ores will be more difficult to reduce).

Mercury does not form strong bonds to anything. It is the only metal that is liquid at room temperature, has quite a low boiling point compared with other metals (357°C), and its vapour consists of single atoms. In fact, mercury vapour was the precedent that Ramsay and Rayleigh (Chapter 5) referred to when arguing from specific heat that argon was also monatomic. Because the bonds between mercury and other atoms are so weak, they are quite easily broken, and we saw in Chapter 3 that oxygen was first isolated simply by heating the oxide of mercury.

Zinc, on the other hand, is a much more reactive metal, so much so that it is on the very borderline of being reducible by carbon. Again, as metals go, it bonds to itself fairly weakly (boiling point 907°C), but it forms reasonably strong bonds to other atoms. Copper forms bonds to other atoms about as strong as those formed by zinc, but bonds to itself very much more strongly, which is why it is unreactive and so easy to prepare from its ores. Iron binds to itself a bit more strongly than copper does, but the bond between iron and oxygen is considerably stronger than that between copper and oxygen, so iron is more difficult to reduce for this reason, as well as for the other reasons (higher melting point, and formation of a range of carbon-containing materials) discussed above. Calcium and magnesium resemble zinc in binding to themselves fairly loosely, and also form very much stronger bonds to oxygen than the other metals we have been discussing, so it is extremely difficult (or in the case of calcium, impossible) to prepare them from their oxides using carbon. Aluminium binds to itself almost as strongly as iron, but its bonds to non-metals are among the strongest that exist, and cannot be broken except by using extremely strong reducing agents (such as sodium metal), or electrochemically. Sodium itself occurs in its compounds as a positively charged ion (see Chapter 12), and is prepared from its chloride electrochemically.

What Next?

Some metals are going to be in increasingly short supply within a very few decades. Indeed some metals, especially the "rare earth" metals used in making powerful magnets, and ultra-strong steel for windmill blades, already are. China is denying reports that it has been cutting off supplies of rare earth elements to Japan, in connection with a sea boundary dispute, while the last US mine to produce these elements is scheduled to restart production in 2012,[52] and deposits in Western Australia and Malaysia are coming on line. Japanese electric car makers are redesigning their motors to use conventional ferrite, rather than rare earth containing, magnets,[53] and the Japanese government is subsidising further research into reducing rare earth use.[54] For the other metals, we can expect ongoing prospecting (although there can by now be very few regions not fully explored), continued research on workup of low-grade ores, and increasing substitution. There is nothing new about the last of these. Aluminium has long since replaced copper for powerlines, although attempts to use it for domestic wiring led to unacceptable problems. The unfortunate cat is no longer on a hot tin roof, but on a hot galvanised roof, because zinc is cheaper than tin by a factor of ten, and just as good at protecting steel. Copper telephone lines are steadily being replaced or bypassed by glass fibre optic or by satellite links. Titanium may be about to become much cheaper, at which stage it will find a host of new applications, and there is no shortage of titanium ore. There will undoubtedly be major pressures on some mineral resources in the coming decades, but other materials will be developed and to some extent replace them. For these reasons, I would not place the exhaustion of metal ores at the forefront of our problems.

To Sum Up

Over half of all elements are metals. They span the entire range from almost the least reactive substances (only outdone by the noble gases) to some of the most reactive, from among the most common to the very rarest elements on Earth, and from substances sought after since Neolithic times to others only isolated in the 20th century. Accordingly, the methods used to isolate them range from the simplest possible (just pick the stuff up) to among the most sophisticated in productive industry.

10 In Praise of Uncertainty

"We may regard the present state of the universe as the effect of its past and the cause of its future. An intellect which at a certain moment would know all forces that set nature in motion, and all positions of all items of which nature is composed, if this intellect were also vast enough to submit these data to analysis, would embrace in a single formula the movements of the greatest bodies of the universe and those of the tiniest atom; for such an intellect nothing would be uncertain and the future just like the past would be present before its eyes."

So wrote the French astronomer and mathematician, Pierre-Simon, Marquis de Laplace, in 1814. Happily, he was wrong from beginning to end.

The Need for Uncertainty

It is usually assumed that uncertainty is a kind of blemish in our knowledge of the world; that we should be trying ideally to remove it. Science in particular is often seen as aiming at an ideal of absolute certainty, where the behaviour of every system, however complex, can in principle be predicted by following the forces acting on all of its components. This kind of determinism is what is known technically as the reductionist point of view, in which the way to understand the whole is to take it to bits and examine the parts.

Laplace has a great deal to answer for. Much hostility to science is actually hostility to Laplace's version of reductionism, on the grounds that it reduces all things to machinery, destroying by analysing, and thereby dissects away the very things that make the world interesting.

We scientists ignore such hostility at our peril, not only because we depend on popular support at every stage of our activities but because these criticisms are largely justified. Our scientific knowledge itself requires us to develop an alternative to the Laplacian view. Laplace's position was based on the search for certainty, by way of reduction. But for over a century, developments within science itself have made this ideal outmoded. Reductionism remains a powerful tool, but only in clearly defined and restricted circumstances.

I will be making five specific claims on behalf of uncertainty

1.   Uncertainty is an essential part of any sane world view
2.   We can learn a lot from the very fact of our own uncertainty
3.   Uncertainty is built into the structure of the physical Universe at its deepest level
4.   If this were not so, we could not have an interesting universe. For example, we could not have a Universe capable of supporting life
5.   The attempt to avoid uncertainty by making predictions is self-defeating in any Universe.

The Philosopher's Paradox

There are many types of uncertainty. The most profoundly inconsequential kind is philosophical uncertainty, at once irrefutable and pointless. How do I know that my perceptions are not all illusions, that I am not dreaming, that the future will be anything like the past etc. Indeed, I may be dreaming without being aware of it, and sometimes I actually am. There are two odd things about these questions. One is that they are utterly unanswerable, if only because of the way they are set up. How can we show that our experience is real, say, without appealing to experience? The other odd thing is that no one really goes around worrying about them. There are philosophers who argue in favour of solipsism, who say they have no good grounds for believing that anything is real other than their own awareness, so that almost all the things we take for granted, such as the reality of a world beyond ourselves, and the fact that other people have minds and feelings, are completely unproven. However, such people are not really as silly as they pretend to be; they are just as good at loving their children and just as careful to insure their lives as the rest of us, and even go to conferences to exchange ideas with colleagues. Such philosophical uncertainty is at once unassailable and absurd. Now I'm not saying that philosophers shouldn't be talking about these questions, but I'm saying they should do so, not to answer them (which can't be done) but in order to exorcise them. Someone who really believed that the world is unreal would be classified as seriously mentally ill. But if the arguments for solipsism are logically so unanswerable, why do we feel so totally justified in dismissing them? The really interesting question, in other words, is not *whether* we are right in ignoring these philosophical uncertainties, but *why* we are right to do so. I shall simply assume that we are indeed right to do so, otherwise there wouldn't be much point in my writing any of this, and pass on.

Not Proven

Much more serious is uncertainty based on lack of evidence, i.e. lack of knowledge that we could have had, but in fact don't. The jury did not see the crime; it is unconvinced; and being uncertain, it returns a verdict of not guilty. Immediately, we run into a complication. Why is the jury unconvinced? For many reasons, but among them surely is the range of expectations that they bring to the case. If you believe, say, that the police never use unnecessary violence, you will believe that the man being pushed to the ground on the videotape really must be posing some kind of threat. On the other hand, if you believe that the police routinely plant and distort evidence, you will be easy to persuade that perhaps a fingerprint was moved, or DNA samples doctored. Obvious enough, but it illustrates one of my central points; that we do not and cannot analyse problems in isolation; that our impressions of certainty or uncertainty over one issue will depend on what we believe about other matters. If we are wise, we will try to take this into account. We call this avoiding bias. But how can we tell what an unbiased view would actually look like? Everything connects with everything else, and our deepest biases are those of which we are the least aware.

How do we respond to uncertainty based on lack of evidence? Not very well. Imagine the fate of a politician, say, who admitted to being

uncertain about how to deal with crime, the long term economic consequences of the budget deficit, or the right way to improve education. Yet
all of these are areas where right now we should be uncertain, so that we
can ask sensible questions, make rational decisions, and leave a reasonable margin for our own errors. After all, no manufacturing company
would invest billions in plant without first going through the best available cost–benefit analysis, but governments spend billions of your money
and mine on prisons, with no evidence that the investment has any value
whatsoever. The US, for all its aversion to spending public money, has
more prisoners per head than any other advanced nation, while the cash-
strapped United Kingdom has more than any other in Western Europe.

There is nothing particularly new about my pointing out the need to
recognise our areas of uncertainty. To quote Jefferson, "Ignorance is preferable to error, and he is less remote from the truth who believes nothing,
than he who believes what is wrong." Or, a century later, Mark Twain,
"It's not what a man doesn't know makes him a fool; it's what he knows
that ain't so." Yet despite the obvious cost to us of our mistakes, we find
it more comfortable to go around looking only for the evidence that confirms our beliefs, rather than what could challenge them. Indeed, when it
comes to our most deeply held beliefs, we often tell each other that to
question them is blasphemous, unpatriotic, disloyal, or in some other way
wicked.

Why this is, I can only speculate, by thinking about the dilemmas that
faced us during our long evolutionary prehistory. One suggestion comes
from thinking about the kinds of threat faced by our ancestors. Is that ripple in the grass a sabre-toothed tiger? That stranger walking towards us
holding a rock, is he planning to throw it? We don't really have time to
wait for further data; by the time we have it, it could well be too late. So
there is strong selection pressure in favour of making up our minds
quickly, and acting decisively. The selection pressure in favour of being
able to review our opinions is much weaker. If we were right, there is no
reason to change our minds. If we were wrong, we may well be dead
anyway.[1]

The other suggestion comes from our social evolution, and was
pointed out by Kurt Vonnegut in *Breakfast of Champions*; no doubt it has
also been pointed out by more "serious" thinkers. We don't only use our
beliefs to help us deal with the world, but also to help us relate to each

other. This second function of beliefs has nothing to do with whether they're true or not, since their only function is to bind groups together, to help us recognise the people who are on our side, the goodies, and to distinguish them from the other lot, who are of course villainous heretics. The evolutionary pressure in favour of culture-based cooperation is potentially even stronger than cooperation based on genetic similarity,[2] since the gang on the other side of the frontier (or, in unhappily divided cities, the other side of the street) may well be quite similar to us genetically, despite being culturally distinct.

Ignorance as a Predictive Tool

So much for the evils of denying uncertainty where uncertainty is what is warranted. So much also for uncertainty as inescapable imperfection, as the unavoidable absence of knowledge. But what I am claiming is much more than this, that the admission of uncertainty can itself be a source of knowledge. This is an idea of the mid to late 19th century, and is associated with the names of such thinkers as James Clerk Maxwell and Ludwig Boltzmann. Think of the air in a room. As you know, it consists of some large number of molecules all rebounding off each other, so large that we have no hope of charting all their motions, which we would need to do to attain Laplace's dream of perfect prediction. However, there is a strange paradox. As soon as we admit our ignorance of the behaviour of the individual particles, we can use this very ignorance as a tool to discover new laws, laws that accurately describe the behaviour of matter in bulk. We have no hope of working out what particular arrangement the air molecules in a room will adopt, but that does not stop us from making very confident predictions about what *kind* of arrangement will be adopted, because some kinds are much more numerous than others. We know nothing at all about where any one molecule is; just as, when a pack of cards is well shuffled, fully randomised, we can make no predictions at all about where any one card is. Then imagine that someone draws four cards at random from the pack. Any arrangement is as good as any other, but if he draws all four aces, you will (to put it mildly) suspect him of cheating. That is not because the hand with four aces is any more unlikely than any other hand (it isn't), but because there is only one hand that has four aces in it, and 270,724 that don't. A powerful statistical argument.

The remarkable thing is, that once we admit ignorance of details and fall back on statistics, we can explain the laws that govern gases. When we have a mixture of different gases, it is extremely probable that all the different kinds of gas will have the same average kinetic energy, spread out around the average in the same kind of way. This leads directly to a justification for Avogadro's hypothesis that different kinds of gases at the same temperature and pressure have the same number of molecules per litre. And this realisation proved (Chapter 4) to be the key to understanding the basics of chemistry. The statistics of how energy is spread out between translation and rotation explains the differences in specific heat between monatomic and polyatomic gases, as in Ramsay and Rayleigh's demonstration that argon was monatomic. And while we do not pretend to know the energy of motion of any one molecule, we can calculate what fraction of collisions between molecules are energetic enough to give rise to chemical reactions.

This surrender to probability is sometimes known as maximising entropy, and the kind of reasoning that is used is known as statistical thermodynamics. This branch of knowledge lays down the limits of the possible in the transformations of heat, work, and chemical energy, and its applications run from the design of heat engines (old fashioned coal-driven steam engines, or the most modern high pressure turbines, it doesn't matter), through the metabolic processes of living things, and on to the birth and death of stars. One application is calculation from heat data of the constants that determine the composition of an equilibrium mixture, as in the case of the production of ammonia from nitrogen and hydrogen, the most important single reaction in the whole of industrial chemistry (Chapter 6). Einstein said of thermodynamics that it was the only scientific theory that he expected would last essentially unchanged for a thousand years; and, let me remind you, the statistical justification of thermodynamics is based on randomness, which is the assumption of ignorance.

A Quantum of Uncertainty

Let us move on to the first quarter of the 20th century; once again science is being driven forward by uncertainty. This new kind of uncertainty is described by the theories of quantum mechanics, and can be summed up by Heisenberg's uncertainty principle. To understand this principle, imagine that you are trying to find a particle under the microscope, and you want to know both where it is now, its position, and how fast it's changing its position. So the first thing you do is to bounce light off it; how else are you going to see it? But light has a wavelength, and you can pin down your particle's position only to an accuracy that is controlled by that wavelength. So, to pin down your particle's position more precisely, you decide to use a shorter wavelength light. Now you're really in trouble, because light acts in packets called photons, and the shorter the wavelength of your light the greater the momentum of each photon, and the more each photon can change how fast the particle is moving. So if you try to pin down position more exactly, you disturb the motion by a larger amount. The upshot is, that if you multiply together the uncertainty in position by the uncertainty in momentum (which is speed times mass), you never, even in principle, get below a certain amount of overall uncertainty, known as Planck's constant (we briefly met Planck himself in Chapter 6, when we discussed the different ways in which German scientists responded to the rise of Hitler).[3] It then turns out, and this is by no means obvious, that the best way to cope with the situation is to think of the particles themselves as having wavelike properties, and the description of the world that results is known as wave mechanics; more on this in the next chapter. This kind of uncertainty also makes it impossible to predict the future, because on the smallest scales at least, when we're talking about particles the size of atoms, the quantum mechanical uncertainty is large enough to make all the difference to their future behaviour. One example of quantum mechanical uncertainty is radioactive decay. The nuclei of uranium, for example, decay at a certain rate, and we can predict, to within statistical error, how many such atoms will decay in, say, the next five minutes, but we cannot in any way tell the difference between the ones that are about to pop off right now and the ones that will prove good for another 4 billion years or so.

Once again, we have the paradox of uncertainty making possible prediction; indeed, life itself depends on the existence of quantum mechanical uncertainty. To see this, consider the behaviour of a hydrogen atom, the simplest atom there is, and a key component of all living things. Without quantum mechanics, every hydrogen atom would be allowed to have an amount of energy just a little bit different from every other hydrogen atom, and that means that its chemistry — the way it can bond to other atoms — would also be a little bit different. Thanks to quantum mechanics, only certain energies are possible. This in turn gives us the situation in which all hydrogen atoms in a given kind of environment have exactly the same properties and show exactly the same behaviour. Because of this, they will form chemical bonds in similar situations of exactly the same strength. If this were not so, chemistry would not be what it is. We would have no guarantee that two and only two hydrogen atoms would stick to one oxygen atom to make water, and certainly no guarantee that the hydrogen atoms which hold in place our enzymes and our nucleic acids would do their job. Without uncertainty, atoms would not obey the laws of quantum mechanics; without quantum mechanics, there would be no such thing as a reliable chemistry; without reliable chemistry, there could be no such thing as life.

According to current physical thinking, the role of uncertainty is even more fundamental. Not only molecules, but the particles that make up the atoms themselves, owe such basic properties as mass to quantum mechanical uncertainty about how well quarks, which are the closest we can get to absolutely fundamental constituents of matter, can be made to fit together.[4]

Sheer Chaos

The most surprising, and to my mind the most exciting, of the uncertainties, was discovered early last century, but it is only in the last few decades that we have had the computing power to explore it. I refer to what physicists call deterministic chaos. A chaotic system does not mean a random system. On the contrary, it is a system where everything would be predictable if we had perfect knowledge, but where the effects on our predictions of any error in our knowledge grow, until at some time out in the future our predictions become totally worthless. Remember that Laplace hoped in principle to predict the future from the position, motion, and forces of every particle in the universe as it is at present. Laplace did not, of course, even imagine knowing these facts to an infinitely large number of decimal places, but he must have assumed that if he had enough decimal places the minute errors remaining would not make any noticeable difference. He was wrong.

Laplace's idea of successful science derived essentially from Newton, who had shown how a simple law of gravity could explain,

with the same exact mathematical framework, the downward acceleration of an apple and the motion of the Moon and planets. Well, up to a
point. If you imagine two stars moving round their common centre of
gravity, a so-called binary system, you can exactly solve Newton's
equations, and it turns out that if you make a very small mistake in
your initial positions or speeds, this just leads to a comparably small
mistake in your predictions — no more. So far, so good. But what if
you have a ternary system, three interacting stars? Such things are less
common than binaries, but do in fact exist. The French mathematician
Henri Poincaré (cousin of Raymond Poincaré, President of France at
the end of World War I) tried to solve exactly this three body problem,
and failed spectacularly. In fact, he didn't just fail, he proved it
couldn't be solved exactly; and worse, he proved that under certain
conditions any change, however small, in the initial conditions would
eventually lead to huge changes in the state of the system. So you have
what has become known technically as *chaos*. It doesn't matter how
well you knew the initial conditions. You can't know them perfectly,
and the error limits on your predictions, however small to start with,
will eventually grow and grow until they become as large as the entire
system.

Still, who cares about ternary stars; let's talk about something more
interesting, like playing snooker or shooting pool. Here again, minute differences become larger and larger. This is because a small error in how
you direct the cue ball becomes a larger and larger error with every successive collision. How small an error could cause a complete change in the
outcome of a game? This problem has been solved for the simplified case
of perfectly smooth and elastic pool balls on a perfectly level table. The
answer is, that the effect of the gravitational pull of a single electron at the
furthest ends of the observable Universe would be more than enough to
make a difference to the outcome. But so what? Can a pool game really
matter all that much? Maybe. Imagine a sequence of events rather like
this. Someone is shooting pool, loses, gets fed up and goes off with friends
to a party and at this party meets the person they eventually marry, have
children with them, these children in turn interact with other people and
the ripple of effects spreads out to change more and more of the future
course of history.

But if you're not convinced that a pool game can change history,
how about that most familiar example of chaos, the weather? The

weather can certainly change history. One example is the storm that destroyed the Spanish Armada sent against England in 1597, helping shift the balance of power in Europe and the pattern of colonisation in the North American continent. Another example is the dreadful winter of 1947, which led to near famine in Europe and contributed to the collapse of democracy in Czechoslovakia, and to the Cold War division of Europe that was to last for over 40 years. But we can't predict the weather more than a few days or maybe weeks ahead. We can predict overall trends for decades ahead, whatever climate change denialists may say. But we cannot say confidently whether next year's summer will be warmer or cooler than this year's. Why not? Because the weather is another example of a chaotic dynamic system, meaning that ever so slight a difference in input will lead if you wait long enough to a major difference in output. Long enough turns out to be at best a few weeks, and we will never be able to make specific predictions much beyond that, however much we know, because what we know can never be enough.

The chaotic nature of the weather was discovered by accident, using what was to become a primary tool for exploring chaos, the computer. Edward Lorenz, a meteorologist at MIT, was trying out some weather modelling equations on his computer. One day, for some reason, he re-checked his calculations, fed in the numbers that the computer had printed out from an earlier run, re-ran the model, and to his surprise got predictions which slowly began to diverge from those he had got first time round. Eventually, he found the cause of the problem. The computer was only printing out the numbers to six decimal places, but during an actual run it was carrying them in its internal memory to eight. It was the difference contained in those last two decimal places that was making the difference to the predictions. The extreme sensitivity of the weather, by the way, is sometimes known as the "butterfly effect", that a butterfly flapping its wings in Brazil can determine whether or not there is a tornado in Texas. Unfortunately, we don't know which butterfly, or when to make it flap. Take the story back a bit further, and you will find, somewhere in the chain of events that led to that butterfly flapping its wings, an example of quantum mechanical uncertainty, perhaps in the exact timing of the decay of one particular radioactive atom, or in whether or not a specific ozone molecule trapped an incoming photon of ultraviolet light. This brings together quantum mechanical and chaotic uncertainty, with the effects of the unpredictable amplified, by way of the unknowable, until the whole world is changed.

The quantum mechanical butterfly effect is built into our very DNA. A single radioactive decay event can trigger a mutation. A single disruptive mutation made Queen Victoria a carrier for haemophilia, eventually through her many offspring undermining half the royal houses of Europe as they moved into the 20th century with its wars and revolutions. A single benign mutation spreading through a species can lead its evolution in a totally new direction, and every living thing now on Earth is the product of a long sequence of such changes.

It Doesn't Compute

Finally, I would mention that computing science gives us more general reasons, independent even of our particular laws of physics, to regard the attempt to eliminate uncertainty as foredoomed. Predictions affect actions, and actions invalidate predictions. Imagine a Universe that contains a machine that tries to predict the behaviour of everything, including the future behaviour of a machine as complicated as itself. It can be shown[5] that such a task is inherently self-defeating. The investment world is precisely such a Universe, with cutting-edge computing technology devoted to anticipating the predictions of equally cutting-edge rivals, and we all know where that's got us.

In Conclusion

The ancient Greeks saw the world as governed by the two opposing principles of chance and necessity. When necessity operated, it left no room for choice. When chance operated, the effects of choice were lost in a string of random consequences. My claim here is that in between the realms of chance and necessity there is the domain of uncertainty, in which everything of interest actually happens. This domain arises in much the same way that between the domain of gases and the domain of crystalline solids there is the domain of liquids. The particles in a liquid do not move independently of each other, but neither are they locked rigidly in one position. Because things interact, the Universe has structure. Because the connections are fluid and complex, the effects of these interactions are, even in principle, unforeseeable. The future is inherently uncertain, and the only computer capable of calculating the future of the Universe is the Universe itself. The smallest difference can make a difference. We cannot predict our lives; we have to live them. We cannot foresee our futures; we have to create them.

11 Everything is Fuzzy

All these 50 years of conscious brooding have brought me no nearer to an answer to the question "What are light quanta?" Nowadays every Tom, Dick and Harry thinks he knows it, but he is mistaken.

Albert Einstein, correspondence with M. Besso, 1951

I think I can safely say that nobody understands quantum mechanics.

Richard Feynman, in The Character of Physical Law (1965) Ch. 6.

Light is Waves

Light travels in waves, and waves do all kinds of exciting things. Go to a pond when the wind blows over it, and watch. The waves spread in every direction. They interact, they reinforce each other where they cross peak-to-peak (*in phase*) and mysteriously cancel out where peak meets trough (*out of phase*). We call this *interference, constructive* where the waves are in phase, and *destructive* when they are out of phase. Anti-noise headphones work by generating sound waves out of phase with the background that the wearer wants to tune out.

Now look at light bouncing off shot silk, a pigeon's neck, or the closely ruled grooves of a CD. Gleaming green in one direction, purple in another. This is also the result of interference, a particular kind called *diffraction*. The feather, the silk, the plastic all have regularly spaced light-reflecting variations. On the CD, the spacing is just the distance between the grooves. In the feather or the silk, the spacing is between variations in thickness in the individual strands or fibres. When the spacing is a bit bigger than the length of the wave, these variations give rise to constructive interference for light bouncing off at some angles but not others. These angles depend on the wavelength of the light compared to the spacing of the grooves.[1] Different colours of light have different wavelengths, and so they are diffracted at different angles.

Only waves can undergo diffraction. So light is waves. Waves of what? Waves of electromagnetic radiation. As Michael Faraday discovered, a changing electric field gives rise to a magnetic field at right angles, and vice versa. So if you have an oscillating electric field in space, jiggling up and down, it will give rise to an oscillating magnetic field jiggling from side to side, which will give rise to an oscillating electric field jiggling up and down ... and so on.

Light comes in all wavelengths. The longest are the radio waves, and as we move to shorter wavelengths through the *spectrum* we go through the microwave region, to the infrared, to the visible (red, orange, yellow, green, blue, indigo,[2] violet in that order), ultraviolet, far ultraviolet, x-rays, and gamma rays. Visible light is of the right wavelength to be diffracted by the regularly repeated variations in a feather. X-rays are of the right wavelength to be diffracted by the fixed distances between the atoms in a crystal, and if we shine x-rays through the crystal, we can work out the arrangement of these atoms from the directions and intensities of the diffracted beams. The first crystal structure determined by x-ray diffraction was that of sodium chloride. The most famous is that of DNA. Here, the experiments were carried out by Rosalind Franklin in Maurice Wilkins's laboratory, and, famously, interpreted by Watson and Crick in 1953. Watson, Crick, and Wilkins shared the 1962 prize for medicine or physiology for this achievement; Rosalind Franklin, tragically, had died of ovarian cancer four years earlier, but not before carrying out the first successful x-ray diffraction of a living or semi-living thing, the tobacco mosaic virus.

Light is Particles

Particles, on the other hand, are relatively boring things. They will keep on going predictably in a straight line, unless they encounter a force, at which point they will simply change speed or direction as the force dictates, and that is all there is to it.

Waves are waves, and particles are particles. Or so it seemed. Throughout the first quarter of the last century, this distinction became progressively blurred and then vanished, and we are still trying to make what sense we can of the consequences.[3]

Energy spreads itself out over many forms, including light. So things glow. This is called *hot body radiation*,[4] although cold bodies do it as well. The colder they are, the dimmer the glow, and the longer the wavelengths. The coldest thing in Nature is outer space, at a temperature of 2.725°C above absolute zero. But even this glows dimly in the microwave region, a pale shadow of the light in which the universe was bathed when it first gave rise to atoms, some 380,000 years after its birth.[5] The charcoal chunks of a barbecue glow red, and the surface of the Sun glows yellow-white hot. The surface of the Earth glows in the infrared, corresponding to an average temperature of around 15°C, and as we shall see later it is the fate of this infrared energy in the atmosphere that is responsible for the greenhouse effect.

Astronomers have a special interest in how temperature affects colour, since they use the wavelength of the light from stars to measure their temperature. So in 1900, Rayleigh, at first alone and then in collaboration with the astronomer James Jeans, set about trying to predict the laws of hot body radiation from the then-known laws of physics. They showed that according to those laws, the emission should keep getting stronger on going to shorter wavelengths. This for the simplest of reasons: there are just that many more ways of fitting a shorter wavelength into a given space. But that's not what happens. The intensity is always greatest at one particular wavelength, and the hotter the body, the shorter that peak wavelength is (think of the shift from red- to yellow- to white-hot). At wavelengths shorter than the peak, the intensity drops, and eventually falls away to zero.

The great German physicist Max Planck (Fritz Haber's boss) approached the problem from the other end, first constructing an equation that would fit the data, and only then thinking about its underlying physical meaning. The equation he found involved the frequency, multiplied by a constant (now always written as h, and called Planck's constant in his honour), divided by temperature. He realised that this would make sense, according to the laws of thermodynamics, only if the energy was being emitted in fixed chunks, or *quanta*, with the size of these chunks proportional to the frequency. (The word was taken directly from Latin: *quantum*, plural *quanta*, original meaning "how much"). In political rhetoric, a "quantum leap" has come to mean a large jump, the very reverse of Planck's meaning, in which a quantum is the very smallest amount possible. And the amounts that we are talking about are indeed very small. Planck's constant is equal to 6.626×10^{-34} J s; that's 0.000 000 000 000 000 000 000 000 000 000 000 6626 J s. Here s stands for second, while J stands for Joule, the unit of energy named in honour of James Prescott Joule, who established the relationship between heat and energy. A Joule, incidentally, is not a very generous measure to start with, being equal to the amount of energy generated in a 1 kW electric kettle in one thousandth of a second.

A shorter wavelength corresponds to a higher frequency (since wavelength and frequency are inversely proportional),[6] and therefore to a larger quantum of light energy. Increasing temperature means that more energy is available, so a greater proportion of the light is emitted in larger quanta, and the peak intensity shifts towards shorter wavelengths.

So light energy is *emitted* in quanta. It follows that, in the reversal of this process, it is *absorbed* in quanta. It was Einstein, at that time still working as an assistant examiner in the Swiss patent office, who showed that it also *acts* in quanta. Planck, like every other physicist since Newton, believed that energy was continuous. So he assumed that his equation applied to just one particular phenomenon, namely hot body radiation. It took Einstein to show that the quanta were real physical entities.

Einstein's responsibilities included examining patent applications for electromagnetic devices. So he would have been extremely familiar with what is called the *photoelectric effect*, used at that time to measure the intensity of light. Metals, by definition, hold on only quite loosely to their outermost electrons, and this is particularly true of the most reactive metals, such as sodium and potassium. Light falling on a film of these metals can cause an electron to be completely ejected, but it was known from experiment that not all light can do this. The frequency of the light has to be above a certain threshold, which depends on the metal. Lower frequency light, however intense, produces no emission, while at any frequency above the threshold the number of electrons ejected is directly proportional to the intensity.

Einstein explained why, if quanta really are separate and unsplittable packets of light energy, this has to be the case. You can only use one quantum at a time.[7] So if the frequency of the light is too low, the energy of that quantum will not be enough to kick an electron out of the metal. Above the threshold, the quantum will have enough energy, plus a surplus. The amount of this surplus will depend on the frequency difference between the light being used, and the threshold. But surplus energy can't just disappear, so what happens to it? Einstein predicted that it would end up as the kinetic energy of the ejected electron. So the higher the frequency above the threshold, the faster the electron will come flying out of the metal.

Einstein published this explanation of the photoelectric effect in 1905. To test his prediction, you need to have a film of highly reactive metal inside a vacuum chamber, together with a way of shining light on it at selected frequencies, and devices for detecting the ejected electrons and measuring their kinetic energies. This was actually achieved in 1915 by R. A. Millikan (now best remembered for the "oil drop experiment" that measured the charge of the electron). He spent several years building what he called "a machine shop *in vacuo*" to do this, and the predicted relationship between frequency and energy was duly confirmed.[8]

It was for this work, rather than for relativity, that Einstein was awarded the Nobel prize in physics, at the surprisingly late date of 1921. By today's lax standards, Einstein was surely worthy of at least two if not three separate prizes; for this work, for Special Relativity and General Relativity, and for his explanation of the random movement (Brownian motion) that finally proved the physical reality of molecules. However, the Committee was considerably more cautious in those days. Max Planck was awarded the 1918 physics prize for his work on the quantum, coinciding with Fritz Haber's chemistry prize, while Millikan received his, for measuring the charge of the electron, in 1923.

Ironically, Planck, Einstein, and Millikan were all hostile to the development of the quantum mechanics in which they played such a large role. As we shall see, quantum mechanics leads remorselessly to indeterminacy, yet Planck said, "The assumption of an absolute determinism is the essential foundation of every scientific enquiry", while Einstein famously remarked that God does not play dice with the world. (To avoid misunderstanding, I should add that Einstein's God was simply his name for the orderly structure of the universe, and that Einstein himself regarded the idea of a personal God who took an active interest in human affairs as infantile.) As for Millikan, he was at first simply incredulous in the face of

Einstein's arguments, built his apparatus in the expectation of *disproving* the energy-frequency prediction, and as late as 1916, after he himself had verified it, still described it as lacking a satisfactory theoretical foundation.

The Crisis in the Atom

In 1911, Rutherford's group published the results of their gold foil experiment, which showed that the atom consisted of a small, dense, positively charged nucleus, and of an almost empty space into which the electrons fitted. The obvious way to think of this is as a kind of miniature solar system, with the electrons orbiting the nucleus, in much the same way that planets orbit the Sun, except that it is electrical and not gravitational attraction that stops the electrons from flying off at a tangent. Rutherford was amazed that most of the atom was almost empty space, but with hindsight perhaps he should not have been. Strictly speaking, the planets do not orbit the Sun, but both Sun and planets orbit a common centre of gravity. It is precisely *because* the planets are so much less massive than the Sun, that their orbits are so much larger, and the same reasoning would apply to the nucleus and the surrounding electrons.

There was, however, another, much more serious, problem. An electric charge moving in a circle should give out energy in the form of electromagnetic radiation (i.e. light). The formulas of classical electrodynamics implied that it should give up all its energy in this way and plummet into the nucleus within the smallest fraction of a second. Obviously, this is not what happens.

Such was the situation when a young man called Neils Bohr arrived in England from Denmark on a postdoctoral travel grant. Bohr[9] had originally intended to spend time at Cambridge with J. J. Thomson, discoverer of the electron, but Thomson was not really interested in the deep theoretical problems that obsessed Bohr. Besides, it probably didn't help that at their first interview, Bohr pointed out an error in Thomson's textbook. However, while at Cambridge, Bohr met Rutherford, and it seems to have been a case of friendship at first sight. Bohr spent time with Rutherford in Manchester, and formed a close relationship that lasted until Rutherford's untimely death 25 years later, at age 66, from a bungled hernia operation.

Bohr was in many ways a complete contrast to Rutherford. Rutherford was the consummate experimentalist, and impatient with vast theoretical concepts. In a letter to Irving Langmuir (whom we will meet in the next chapter) he described himself as "a great believer in the simplicity of things" and is reputed to have said that he didn't want to hear anyone talking about "the Universe" in his laboratory. Bohr, on the other hand, was in the deepest sense a theoretician, remarkable not so much for any particular mathematical skill as for his intellectual courage, and his deep intuitive understanding of where ideas were leading. He was as much a philosopher as a scientist, and spent a great deal of his time thinking about the Universe (or at least, about what we can know about it). His Ph.D. topic had been on how electrons conduct heat and carry current in metals. A seemingly prosaic topic, but it gave him critical insights into the shortcomings of the theories then available.

Bohr wrote to Rutherford in 1912, saying that the stability of atoms required a new hypothesis "for which there will be given no attempt at a mechanical foundation (as it seems hopeless)". Bohr did himself far less than justice with these words. His new hypothesis had no basis in traditional mechanics, but it was the mechanics, not the hypothesis, that was at fault.

Classical physics couldn't explain hot body radiation at short wavelengths. Planck, as we have seen, got around the problem by supposing that light energy could only be generated in chunks, equal to h times the frequency of light. Another thing classical physics couldn't explain was the specific heat of solids (the amount of heat energy required to raise the temperature of one gram of a material by 1°C). As we saw in Chapter 4, this generally obeys the rule of Dulong and Petit at room temperature, according to which the specific heat multiplied by the atomic weight is the same for all solids. This is as expected from the classical statistics of heat energy. However, specific heats actually get smaller at lower temperatures, approaching zero as the temperature approaches absolute zero. This was no mere curiosity, but part of the explanation of Nernst's "heat theorem", the Third Law of Thermodynamics, which decides the favoured outcome of chemical reactions. Einstein explained the loss of specific heat at low temperatures in 1907,[10] by suggesting that heat energy also came in chunks, equal to h times the frequency with which atoms could vibrate around their most stable positions in a crystal. Bohr was certainly very familiar with this work, and in 1913 came up with his own equation involving energy, h, and frequency, and applied it to the hydrogen atom.[11] This is the simplest possible system. It consists of a positively charged nucleus, containing nearly all the mass of the atom, and a negatively charged electron. Bohr started from Rutherford's model of the atom, in which the centrifugal force on the electron is balanced by the electrical attraction of the nucleus, and worked out what would happen if the electron had to obey this classical requirement, and the proposed new equation, at the same time.

In a series of papers published in 1913,[12] Bohr showed that only some specific orbits could meet these requirements. These orbits formed a series with definite energies, like rungs on a ladder. The lowest rung corresponds to the most stable state possible for a hydrogen atom. The rungs above this form a series, getting closer together in energy as they approach a limit, the *ionisation energy*, which is the energy needed to completely detach the electron from the atom. An electron can absorb or emit light by jumping from one rung to another, and one quantum of that light has to correspond to the difference in energy between those rungs. This explains why the hydrogen atom can only absorb or emit light at some particular frequencies. We say that it shows a *line spectrum*, in contrast to the continuous emission from a hot body. The lines in the spectrum of hydrogen had already been measured and tabulated in great detail, and their frequencies, multiplied (of course) by Planck's constant, were an exact match to the energy differences calculated by applying Bohr's rule to the different rungs of the ladder.

Bohr next considered many-electron atoms, and proposed, correctly as we now know, that here also the line spectra showed that only some energies were allowed. Finally, Bohr suggested that in the hydrogen molecule, H_2, the electrons are shared between the two atoms, rotating around a point halfway between the two atomic nuclei. Bohr's specific model is now of historical interest only, but the concept of electron sharing remains central to our picture of chemical bonding. More on this below, and in the next chapter.

Bohr's work was at once a triumph, and deeply unsatisfactory. It explained why atoms, in general, can absorb or emit light of some frequencies but not others, and why each element has its own characteristic set of frequencies, one of the facts that Rayleigh and Ramsay had used to show that argon was indeed a new element. For the simple case of the hydrogen atom, it gave the actual values of these frequencies. It led to the prediction that a helium atom that had lost one of its two electrons would show the same frequencies as a hydrogen atom, multiplied by a factor very close to 4, since in Bohr's model, the frequency for a one-electron system should be proportional to the *square* of the charge on the nucleus. Finally, and this is the aspect that drew Einstein's admiration, that factor should not be exactly 4 but, in agreement with experiment, 4.0016, once we take into account the fact that the nucleus, as well as the electron, is rotating around the atom's centre of mass.[13]

In using the frequency of light as a measure of energy differences, Bohr was simply following the lead already given by Planck and Einstein. On the other hand, and this is something of which Bohr was very well aware, his central assumption limiting the range of possible orbits was a completely novel rule, tacked onto classical physics to make it fit the facts.

Indeterminacy

At this stage, we have three separate demonstrations of the quantum nature of light; Planck's theory of hot body radiation, Einstein's explanation of the photoelectric effect, and Bohr's use of light frequency as a measure of the differences in energy between the possible states of an atom. But why should this lead to *indeterminacy*, a fundamental lack of certainty about what is going on at the atomic scale? To see this, we need to go back in time to the *double slit experiment*, a version of which[14] was first carried out by the Englishman Thomas Young in 1801. Take a beam of light, shine it through a narrow slit onto a screen, and you will see a band of light on the screen. Of course; light travels in straight lines, as has been known since ancient times. Now, instead of one slit, take two slits very close together. If, as Newton thought, light consisted of particles, you would expect to just see two bands of light side by side. This is not what happens. You actually see bands of colour, and if the light itself consists of only one frequency, like the yellow-orange light from a sodium discharge lamp, you see alternating dark and bright bands. This is the simplest possible example of diffraction; at some positions on the viewing screen, light from the two different slits will be in phase and reinforcing, while at other positions the two different pathways will be out of phase and cancel each other out.

Now replace the viewing screen by a photographic plate, and you will get the same kind of diffracted image. No surprises so far.

Now reduce the intensity of the light so much, that only one quantum is emitted at a time. That quantum will hit the photographic plate, and be recorded there. Wait long enough, and a whole host of quanta will have hit the plate, each of them having arrived quite separately from the others. So, you might think, each quantum must have gone either through the left-hand slit, or through the right-hand slit, and now, surely, you should get two separate bands, one from each slit.

You would be wrong.

Each separate quantum of light has *acted* as a particle, at one specific point on the photographic plate. But the *probability* of it doing so at any point depends on the intensity of a wave. Did this or that particular quantum go through the left-hand slit or the right-hand slit? The answer to this question is *indeterminate*. There is one possible history in which it went through the one, an equally possible history in which it went through the other, and the way in which these possibilities combine follows the arithmetic of interfering wave patterns. The light quantum itself doesn't know which slit it went through, or if it does, it's not telling.

In short, each individual *photon*, as the quantum of light is now called, travels as a wave, but acts as a particle, and it is the intensity of that wave that determines the probability of it acting at any one specific point. There is no saying where the photon *really is*, until it acts, and even then, there is no saying which slit it went through, so there is no saying where it *really was* before it acted.

If you find this description of the world completely unsatisfactory, then you are, as we have seen, in excellent company.

Particles are Waves, But No One Understands How This is Possible

It gets worse. Physicists are always searching for symmetries in Nature. If light, while travelling as a wave, is emitted and absorbed in particle-like quanta, could it be that ordinary particles might show some of the properties of waves? And does that include the property of being spread out in space? And if so, would that not introduce intrinsic uncertainty into the behaviour of all matter at very small scales? Yes, to all these questions, according to the "Copenhagen interpretation" of quantum mechanics developed by Niels Bohr and accepted by most (not all) physicists to this day.[15]

There are yet more serious paradoxes, involving what is called *entanglement*. Some processes can give rise to two particles (or light quanta) in a single step. When this happens, a measurement on one member of the pair seems to immediately affect the results of a similar measurement on the other, giving apparently instantaneous communication, which is of course impossible, but that is another long story.[16] Finally, and to my mind worst of all, there is a paradox about the very act of measurement. The state of an object we are looking at can be indeterminate until we carry out the relevant measurement. But the measuring apparatus itself is subject to the laws of quantum mechanics, and its own state is therefore uncertain until observed. Does a grain on a photographic plate count as an observer, or does the observer have to be conscious? How conscious? Does an ant have enough consciousness, or Schrödinger's famous cat,[17] or do we need a human being? Why such a special role for consciousness, and is this not merely replacing one mystery with another? In any case, if consciousness is the key, was the entire universe indeterminate until the first conscious beings evolved? Some people (I hasten to say none of those mentioned here) have actually suggested as much.

At this point, all I can do is quote Richard Feynman, who did after all win the Nobel prize for his work on the subject: "I think I can safely say that nobody understands quantum mechanics", and, later, "Do not keep saying to yourself, if you can possibly avoid it, 'But how can it be like that?' because you will get 'down the drain,' into a blind alley from which nobody has yet escaped. Nobody knows how it can be like that." Nobody knows, but quantum mechanics explains the behaviour of light, and of the smallest particles of matter, in exquisite detail. We may not know what it means, but we know what it says when it comes to working out how electrons (or other things) behave; how there are restrictions on how the electron "wave" can fit inside the atom, how, as a result, only some energy states for electrons are possible, how the sharing of electrons between atoms gives rise to the chemical bonds that hold molecules together, how and why this gives molecules their definite shape, and materials their characteristic colours, and how all of this makes life possible.

The matching of wave properties to particles, mirroring the particle-like properties of light waves, was proposed in a Ph.D. thesis by Louis de Broglie in 1924.[18] His examiners, understandably unsure what to make of this idea, sent the thesis to Einstein for review. Einstein approved. So, in 1929, did the Nobel Physics Committee (5 years from Ph.D. to Nobel Prize. Wow!)

We can readily put de Broglie's ideas to the test. De Broglie's proposal was that the momentum of a particle (mass times velocity), multiplied by the wavelength, should be equal to h. A little calculation shows that accelerating electrons through 100,000 volts will give each of them the correct momentum for their wavelengths to be comparable to the distances between atoms in a crystal. Such electrons should undergo diffraction, much as x-rays are diffracted. Moreover, it should be possible to obtain these diffraction patterns using much smaller samples than those required for x-ray crystallography, since electrons interact with matter much more strongly than x-rays. This is indeed observed, and electron diffraction is now among the standard methods for determining the crystal structure of solids.

De Broglie's approach provides the missing justification for Bohr's assumption. Stop thinking of the electron in the atom as a particle, and think of it instead as a wave. This wave needs to join up with itself as the electron travels around an orbit. So the orbit needs to be a whole number of wavelengths. And, as always, electrostatic attraction and centrifugal force need to balance. But centrifugal force depends on momentum, and according to de Broglie, the momentum determines the wavelength. Combine all these requirements, and you get the same set of allowed energy levels as in Bohr's model, but a very different interpretation. We have lost the idea of the electron as a particle with a definite position, and replaced it with the idea of a wave spread out in space. There is no such thing any more as a definite position for the particle. We have, at best, a probability of finding the particle close to any particular point, corresponding to the intensity of the wave at that location.[19] I have cheated a little here, treating the wave function as constrained to a single loop whereas in reality it is spread out over a spherically symmetrical three-dimensional space, but while the mathematics are more complicated the principle is the same.

From here on in, we will no longer speak of the "orbit" of an electron, since that suggests a well-defined position, but of its orbital. We have entered a new era in our attempt to understand the physical world, the era of wave mechanics.

Many people contributed to the development of wave mechanics, but two names stand out, those of Erwin Schrödinger and Werner Heisenberg. In the next paragraph, I shall attempt to explain just what they did, and what the difference was between their two approaches. However, if you don't want to get that technical, you can skip without affecting what comes later.

Schrödinger took the concept of a "wave" very seriously, and developed a set of what are called partial differential equations to describe the wave-like behaviour of matter. (If you don't know what a partial differential equation is, don't let that bother you. Not all that many people do. All you need to know is that they can be used to describe the behaviour of waves.) Heisenberg developed what looked like a completely different mathematical treatment, in which measurements were represented by mathematical operations, but very soon Schrödinger showed that the two treatments are, at root, identical. Heisenberg's famous uncertainty principle says in effect that what we learn about one aspect of an object (for example, its momentum) affects how much we can know about a different aspect (in this example, its position). This turns out to correspond to the fact that the mathematical operations corresponding to the two different measurements do not, as mathematicians say, commute; it depends which order you do them in. A trivial example of not commuting is adding three and doubling. Take seven, add three, and double, and you get 20; take seven, double, add three, and you get 17. The mathematical operations of wave mechanics involve rather more complicated things, such as partial differentiation by x before or after multiplying by x, but the same principle applies. In Heisenberg's formulation, the non-commuting operators are represented by non-commuting matrices. If that last sentence doesn't mean anything to you, don't worry. It wouldn't have meant anything to most physicists before 1927.

So far, I have been talking about a single electron in a single atom. You may be wondering, if electrons are wave-like, whether they can spread out over more than one atom, and, if so, what can happen as a result. The answer is yes, that when they do so they can be attracted to more than one nucleus at the same time, and that this is the solution to the problem that had puzzled chemists since Avogadro; what kind of force could be holding the atoms of a molecule together, if the atoms are both of the same kind. In order to see how this takes place, and what happens as a result, we need to discuss how many electrons can fit into the same orbital, what kinds of rules decide what orbitals are available, and what follows from the fact that electrons repel each other. With this information, we will be able to understand why some molecules are stable, and why they have the shape they do. But this forms the subject of the next chapter.

You may also be wondering if the uncertainty principle applies to the nuclei of atoms, as well as their electrons. It does indeed.

Think about any two atoms that are bonded together. There will be an equilibrium distance, at which the bond is most stable. In classical physics, the atoms would just settle down at this distance, especially at low temperature when the molecule has little energy. Thanks to the uncertainty principle, this can't happen. For if the atoms were at their equilibrium distance, and stationary, then we would know at one and the same time their position relative to each other (the equilibrium distance), and their momentum (zero). This would violate the uncertainty principle. It follows that even at absolute zero, every bond has a certain amount of energy that it can't get rid of, the *zero point energy*. Zero point energy turns out to depend on the masses of the atoms, as well as on the strengths of the forces between them. So it is responsible for small quantitative differences in chemical behaviour between different isotopes (different mass nuclei) of the same element. This gives rise among other things to the temperature-dependent effects that are used to study past climate, as explained in Chapter 8. It can also give important information about the past processing of materials, which is one of the ways in which we know that the carbon dioxide added to the atmosphere in the past 150 years comes from the burning of fossil fuel, rather than, for example, the decomposition of limestone. The application of quantum mechanics to nuclei bonded together also explains why some (not all) molecules absorb infrared light, giving rise to the greenhouse effect discussed in Chapter 15.

Can we diffract whole atoms? Yes. An early 2011 study[20] even used a diffracted beam of alkali metal atoms to study how strongly they interacted with the slit material.

What Happened to Everyone

Schrödinger (Nobel prize 1933) became professor of physics at the Friedrich Wilhelm University in Berlin, now Humboldt University. In 1933, he chose to leave Germany, although as an "Aryan" he was under no personal pressure, and spent time at Oxford and Princeton before taking up a professorship in 1936 at the University of Graz in Austria. Bad move. In 1939, the Nazis moved in, and he had to move on, dodging across the border into Switzerland and eventually taking up a permanent position in Dublin, accompanied, as always, by his wife and his mistress. In 1943, he delivered a series of lectures on what we would now call the informational aspects of biology, published the next year as *What is Life?* a short book with enormous influence.

Niels Bohr became professor of physics at Copenhagen in 1916, aged 31, and in 1921 established the Institute for Theoretical Physics, a major centre of influence in the development of quantum mechanics. The following year, he was awarded the Nobel prize. He was responsible for the idea that electrons in atoms formed "closed shells", and that only those outside the shells took part in chemical bonding. He was also responsible for the realisation, in 1939, that it was specifically uranium-235, and not the much more abundant uranium-238, that was involved in nuclear fission.

In 1940, the Germans occupied Denmark, but imposed Protectorate status rather than annexing it, and it was not until late 1943 that Hitler ordered the roundup of all Danish Jews. Acting on an inside tip-off,[21] Danish fishermen forestalled this action by transporting almost the entire Jewish population across the straits to neutral Sweden. Niels Bohr, whose mother was Jewish, was among those rescued. Bohr was taken to London and thence to the United States, where he worked at Los Alamos on the Manhattan Project for the development of the atomic bomb. He attempted to persuade both Churchill and Roosevelt to share their nuclear secrets with the Soviets, but they were not impressed by his arguments.

De Broglie[22] was appointed Professor of theoretical physics at the University of Paris in 1932. He stayed in occupied Paris throughout World War II, and was elected to the Académie Française in December 1944, when it reconvened a few months after liberation. He ran courses at the Institut Henri Poincaré, was one of the many founders of quantum mechanics (along with Planck, Einstein, and Millikan) to attempt to over-throw its probabilistic interpretation, wrote prolifically both for his scientific colleagues and for a wider audience, and was honoured by UNESCO for his work in explaining science to the general public.

Heisenberg (Nobel prize 1932) was a member of Bohr's group in Copenhagen, where his version of quantum mechanics led him to the uncertainty principle that bears his name, before returning to Germany in 1927. As a theoretical physicist, he fell under suspicion when the Nazis came to power, but successfully appealed to Himmler for protection. During World War II, he was heavily involved with the German nuclear programme. He visited Bohr in 1941, and the subject of their conversation (which Bohr abruptly terminated) has been the subject of much speculation. Be that as it may, in 1942 he advised the German government that there was no prospect of Germany building a nuclear weapon in time to influence the outcome of the war, and German research in that area was consequently downgraded. As to whether that advice was influenced by concern about a nuclear weapon in Nazi hands, we must remain uncertain.

To Sum Up

Light consists of waves; waves of electromagnetic radiation, of disturbance in electrical and magnetic fields in space. Yet light is generated, and acts, in quanta, called photons, and these photons can be considered as particles. This paradox leads to *indeterminacy*, since a wave, by definition, is spread out in space. But particles also show wave-like behaviour, so that indeterminacy is built into the behaviour of matter as well as light. By applying constraints derived from wave mechanics to electrons, we can make sense of the stability of atoms, and of their absorption and emission of light. We can also understand, as we shall see in the next few chapters, how atoms bond together, why molecules have definite shapes, and how they recognise each other in biological systems, making life as we know it possible.

12 Why Things Have Shapes

If everything is made of atoms, and molecules are made of atoms stuck together, what are the rules that govern how many of each kind stick to each other, and how they arrange themselves in three-dimensional space when they do so?

Wasted lives

If I should die, think only this of me:
That there's some corner of a foreign field
that is for ever England.

Rupert Brooke, 1887–1915

By late 1914, the World War I Western Front had reached a stalemate, that would persist for four more murderous years. Germany and Austria-Hungary controlled the Baltic and the land routes from the West to Russia. The Allies tried to bribe the Ottoman Empire into entering the war on their side, but the Germans simply offered a larger bribe, which was accepted. This meant that the Allies could no longer supply Russia through the Dardanelles, the narrow straits that link the Mediterranean with the Black Sea and Russia's only warm water ports.

Winston Churchill, then First Lord of the Admiralty, had the bright idea of using obsolete British battleships to gain control of the Dardanelles. The original plan was for a purely naval operation, with ships forcing their way through the straits, which are at one point only a mile wide. Unfortunately, the Germans had mined these waters, while the Turks were able to bombard ships in the straits from the neighbouring high ground. The Allied naval forces suffered unacceptable losses, and withdrew.

To the British high command, the next move was obvious. Send troops ashore to capture the strategic high ground of the Gallipoli peninsula, and with the Turkish guns silenced, minesweepers could reopen the straits. Unfortunately, it took several weeks to put this plan into operation, during which the Turks were able to dig in and reinforce their positions. The Allies established themselves on the beaches, but despite inflicting (and suffering) horrendous casualties, failed in numerous attempts to capture the heights. They withdrew in December 1915, a year after the initial naval assault and eight months after the first landings, with nothing accomplished.

Rupert Brooke died from an infected mosquito bite on the ship taking him to Gallipoli. Over 80,000 Turkish and other Ottoman soldiers died at the front, and more than 44,000 on the Allied side.

Chemistry by the Numbers

As we have seen, Lavoisier gave us our modern concept of an element. Dalton proposed that elements and their compounds were made up of atoms. Avogadro's hypothesis, the law of Dulong and Petit, and extensive chemical analyses led to the building up of a table of atomic weights,[1] and by the late 19th century scientists in several countries were looking at relationships between these weights and chemical properties. The most ambitious and successful of these was Mendeleev in Russia, and the Periodic Table[2] that we use today is a direct descendant of the one that he proposed in 1869. The only major subsequent modification to that Table came, as we saw in Chapter 5, with the discovery of the noble gases in the 1890s. Apart from that, and a bit of confusion about the lightest elements (see below), the job was one of filling in the gaps. From the 1940s onwards, we have been extending the table beyond uranium by bombarding the heaviest elements with neutrons and, more recently, with the nuclei of lighter atoms. The first few transuranic (beyond uranium) elements, at least, behave exactly as Mendeleev's arguments would lead us to predict, although beyond them our samples are too small and too short-lived for us to be able to examine their chemical behaviour.

The Table as originally put forward involved listing the elements in order of increasing atomic weight. Do this, and you notice some regular patterns, and these patterns became even more obvious as more elements were discovered. For example, every so often you come across a noble gas. When you do, the element directly before the noble gas is[3] a reactive volatile element such as chlorine, while the element directly after the noble gas is a soft, low melting, reactive metal such as sodium (Na), which combines with chlorine to make a 1:1 salt such as sodium chloride, NaCl. This is in turn followed by a harder metal such as magnesium that makes a 1:2 salt, $MgCl_2$, with chlorine and a 1:1 compound, MgO, with oxygen, and so on. The regular recurrence of these patterns is what gives the Periodic Table its name, and leads to the arrangement of elements in groups (the noble gases form one group; the elements lithium, sodium, potassium ... directly following a noble gas form another, the chemically similar elements beryllium, magnesium, calcium ... the next group, and so on).

This arrangement into groups is enormously successful in describing trends and regularities in chemical behaviour. However, there is something rather puzzling about the exact relationship involved. Early in the Table, elements are generally separated by around two or occasionally one or three mass units, with larger separations becoming progressively more common as we move to heavier elements. If atomic weight really is the controlling factor, we would expect all the gaps to be the same. Even worse, there are places where the original listing by weight gave what is obviously the "wrong" order, when compared with arrangement by chemical properties. Two such anomalies known to Mendeleev were cobalt/nickel and tellurium/iodine, while subsequent discoveries gave argon/potassium and thorium/protactinium. With hindsight, the reason is obvious, as it so often is. Atomic weight itself cannot be what determines chemical behaviour. Something else must be responsible, something distinct from weight but strongly related to it.

In 1911, Geiger and Marsden, working in Rutherford's laboratory, showed that the atom consisted of a small, massive, positively charged nucleus, surrounded by electrons. Further work showed that the amount of positive charge on the nucleus, compared with the charge on a hydrogen nucleus, was roughly half the atomic weight. Barkla in Liverpool, looking at the scattering of x-rays, got the same answer for the number of electrons. This similarity is what we would expect, since the positive charge on the nucleus must be equal to the total negative charge on the electrons. Interestingly, he paid particular attention to the case of cobalt and nickel, showing that nickel scattered the x-rays more effectively than cobalt, despite its lower atomic weight.

The next logical step was taken by a Dutch lawyer called Antonius van den Broek, self-educated in science, who suggested that the number as determined by rank in the Periodic Table, and the number of units of charge on the atomic nucleus, are *one and the same thing*, for which Rutherford coined the name *atomic number*. As to why the atomic weight is around twice the atomic number, it was assumed at the time that the nucleus contained both positive and negative charges. It was not until 1921 that Rutherford would suggest the existence of an uncharged particle in the nucleus, which he called the *neutron*, as distinct from the positively charged hydrogen nucleus or *proton*. For the lighter elements, the number of neutrons is roughly equal to the number of protons, but the ratio of neutrons to protons slowly increases with atomic number, and is over 1.5 to 1 by the time we reach uranium.

The most direct evidence for the fundamental importance of atomic number came from Rutherford's student at Manchester, Henry Moseley, in a series of elegantly reasoned papers in 1913–1914.[4] Moseley's findings were so influential that many authors give him the credit for inventing the atomic number concept, although his papers clearly acknowledge that it came from Broek. It was already known that solids bombarded with fast moving electrons gave rise to x-rays, and that these x-rays are diffracted by crystals. The angle of diffraction depends in well understood ways on the wavelength, and from the wavelength it is easy enough to calculate the frequency, and hence the energy, of the photons. Moseley found that this energy increased with the chemically defined atomic number, rather than with atomic weight, since it was greater for nickel than for cobalt. Moreover, the actual energies could be fitted into a relatively simple formula, which bore a striking resemblance to the one that Bohr had derived for the spectral lines of hydrogen.[5] Moseley regarded this as direct confirmation of Broek's suggestion, examined a large number of different elements, correctly located the gaps then still remaining in the Periodic Table, and was even able to show that what was thought to be a single rare earth element was actually a mixture of two. As a bonus, Moseley also noticed that some of his specimens of supposedly pure material gave faint impurity signals as well as those for the main component, and correctly predicted that this could be useful as a method of chemical analysis. Rutherford — who once said that qualitative is bad quantitative — must have been utterly delighted by this quantitative explanation of the atomic number concept, originally derived from the qualitative data of chemistry, in terms of nuclear charge.

The outbreak of war unleashed, in all the countries involved, a wave of naive patriotic fervour that is difficult for us now to understand. Moseley had planned to move on from Manchester to a position in Oxford, but now he interrupted his career to enlist. He was sent as a signals engineer to Gallipoli where, in the act of telephoning an order, he was shot through the head by a Turkish sniper.

After Moseley's death, the British belatedly followed the German example, and stopped allowing talented scientists to enlist for front-line duty. Churchill, as initiator of the entire disastrous operation at Gallipoli, lost his position at the Admiralty, but remained in the Cabinet.

Particles in Motion

"If there's a promise held out, it must be honored. Whatever is hidden behind the curtain must be revealed at last, and it must be at one and the same time completely unexpected and inevitable." So said the author and critic Margaret Atwood, speaking of literature. Remarkably, the same applies to the greatest scientific discoveries. And so it came about that, unexpectedly but inevitably, physicists thinking of the smallest fundamental particles, and chemists manipulating entire molecules, slowly converged in the early 20th century on a common vision of the forces that hold atoms together and make our everyday material world possible.

The most reactive metals (see Chapter 9) have to be produced using electrical, rather than chemical, energy. Chemical methods, when they work, are cheaper, but the less reactive metals could also be produced electrically if we wanted to. This led Davy and Faraday early in the 19th century to suggest that chemical bonding was basically electrical in nature. But, as several scientists throughout that century speculated, if electricity can cause chemical change, and if chemical changes involve atoms, then there must be such a thing as an "atom" of electricity. So when J. J. Thomson, in 1897, identified such particles with the "cathode rays" emitted by negatively charged hot metal wires, the name "electron" was already waiting.

Pure water is a very poor conductor of electricity. So is sugar water. But dissolve a salt, such as common table salt, sodium chloride, and the water becomes a good conductor, because the sodium chloride solution contains electrically conducting particles. This much was known to Davy and Faraday, who thought that electrical energy was necessary to generate these mobile charged particles. Much later Svante Arrhenius in Sweden and Jacobus van't Hoff in the Netherlands realised that the charged particles, *ions* as Faraday had named them, were already present in solution and moving at random, more or less independently of each other. With hindsight, the experiment that shows this is a remarkably simple one. If you dissolve anything in water, you lower the freezing point, because removing ice (pure solid water) from a solution is that much more difficult, involving a greater decrease in entropy, than removing ice from pure liquid water. It turns out that, "molecule for molecule", table salt, NaCl, is almost twice as effective as sugar. Salt is also more effective, in the same ratio, than sugar when it comes to generating osmotic pressure. The reason is, that the "molecule" of sodium chloride is completely separated in the solution, into independently mobile[6] sodium *cations* (positive ions), Na^+, and chloride *anions* (negative ions), Cl^-. Later, as mentioned in the preceding chapter, x-rays would be used to find out how sodium and chlorine held together in the solid. It was found that each sodium cation is surrounded by six chloride anions, and each chloride by six sodium cations (remember that opposite charges attract, whereas charges of the same sign repel each other). There is no such thing as a sodium chloride "molecule".[7] During electrolysis of molten sodium chloride, the ions are attracted towards the electrode of opposite charge. At the negative electrode, an electron is forced on to the sodium cations, converting them to electrically neutral metal. At the positive electrode, electrons are removed from the chloride anions, and the resulting neutral atoms link up to form the molecules of Cl_2 that make up chlorine gas.

It may be no accident that van't Hoff and Arrhenius were both working outside Germany, which at that time led the world in chemistry as in other branches of science. Both of them were challenging orthodox doctrine, both began to do so while still graduate students, and both had to face initial opposition and ridicule. These might well have been even stronger in a country that had its own established way of doing things. Likewise, Niels Bohr was to describe himself as fortunate to have come from a small country, with no distinctive scientific style of its own, leaving him more free as a result to pick and choose between different intellectual traditions.

By the time of Nobel's death, Arrhenius was the most powerful force in Swedish science, and dominated the affairs of the Nobel committee, especially regarding chemistry, for over a decade. Arrhenius, who initially saw Nernst as an ally, quarrelled with him (one of a long list of people to do so, including, as we have seen, Haber), and was probably responsible for the long delay before Nernst's heat theorem was recognised with a Nobel Prize.[8]

The Rule of Eight

Some people are very easy to quarrel with. They may take offence too readily, or, having been offended, nurse their resentment unreasonably long, or themselves give offence by arrogant dismissal of ideas that are not to their liking. Gilbert N. Lewis was guilty of all of these. He was initially contemptuous of Bohr's accomplishments and of quantum physics in general, saying in his most important single paper[9] that along such lines "it seems hardly likely that much progress can be made". Despite which, he was among the most important scientists of his generation, and clearly the greatest chemist since its establishment not to win the Nobel Prize, for which he was nominated 35 times.

Theodore Richards, Lewis's Ph.D. advisor, was extremely thorough, and extremely dull. Most of his scientific work was spent determining atomic weights, by the classical methods of using the balance to find out exactly how much of one substance could be produced by complete reaction of another. For this, he was awarded the Nobel Chemistry Prize for 1914, not announced (because of the war) until 1919. By the time of that announcement, it was already well established that many elements were mixtures of different isotopes, with different atomic weights, whose average had no fundamental significance whatsoever. Richards himself helped show this;[10] the atomic weight of lead could vary by more than a unit, depending on where it came from. We can now readily understand this in terms of radioactive decay; the long-lived uranium and thorium precursors ^{238}U, ^{235}U, and ^{232}Th eventually give the stable lead isotopes ^{206}Pb, ^{207}Pb, and ^{208}Pb respectively.

In addition to atomic weights, Richards was interested in thermodynamics, specialising in the measurement of heats of reaction. However, his appreciation of the mathematical aspects of the topic was poor, and G. N. Lewis, while still his Ph.D. student at Harvard, had far surpassed him. Richards's appreciation of his own graduate students was sadly limited. He believed that they needed to be carefully superintended, and that "the more brilliant ones often strike out on blind paths of their own if not carefully watched."[11] And so it came about that when Lewis published his formulation of thermodynamics, Richards incorrectly imagined, because of some similarities in terminology, that Lewis had failed to acknowledge his contributions, while Lewis saw Richards as attempting to steal from his own graduate student, and was still muttering about the matter 40 years later. Lewis himself, when in due course he had his own department, gave the graduate students there the greatest possible freedom.

This was not the only priority dispute that Richards was involved in. He had compiled detailed data regarding the heats of chemical reactions, and also what is known as their "free energy", the maximum amount of work that could be obtained from them. Richards noted that heat and free energy tended to reach the same value as one approached absolute zero, but missed the much more interesting fact that they also tended towards the same slope, like curves touching instead of merely meeting. If he had discussed the data with Lewis, Lewis would assuredly have pointed this out to him, and between them it is likely that with this information they would have anticipated Nernst's Heat Theorem (see Chapter 6), and with it the ability to predict the equilibrium behaviour of chemical reactions.[12] Richards ended up believing that both Lewis and Nernst had stolen from him, but this undoubtedly sincere belief came from his own failure to understand the mathematical subtleties involved.

Having finished his Ph.D., Lewis visited Germany, an almost obligatory step at that time for an aspiring American chemist, and spent some time in Nernst's laboratories, where they laid the foundations for a lifelong enmity. Lewis returned to Harvard as an instructor, but left after three years for a government position in the Philippines, which the United States had recently acquired as a possession, by displacing the Spanish and then suppressing the islands' own independent government. A couple of years later, he returned to Cambridge, Mass., this time to MIT, before becoming Professor and Dean of the College of Chemistry at the University of California, Berkeley, which he helped build up into one of the world's leading research institutions.

While still teaching at Harvard, Lewis developed his own, rather eccentric, model of the atom, a curious mixture of deep insights, and of geometric notions that now seem merely fanciful. According to this, each atom consists of a core (Lewis called this the kernel), and an outer shell of electrons. These *valence electrons*, as they are now called, sit at the corners of a cube. The ideal arrangement is when the cube is complete, so the ideal number of outer electrons is eight. The core itself consists of one or more concentric layers, or cubes, of electrons, although the innermost part of the core, corresponding to helium, is[13] unique and contains only two electrons. The number eight derives from chemical knowledge, and in particular from observations made by Abegg (Clara Immerwahr Haber's adviser), whom Lewis acknowledges, about the bonding abilities of different kinds of atom.[14]

Why a cube? To quote Lewis, "Perhaps the chief reasons for assuming the cubical structure were [were, past tense, because he is describing in 1916 what he tells us he wrote down in 1902] that this is the most symmetrical arrangement of eight electrons, and is the one in which the electrons are farthest apart." Actually, that last statement isn't even true, since twisting the top of the cube through 45° will increase the distance between the top and bottom electrons. Why a cube with its eight corners rather than, say, a regular 6-cornered octahedron, which is just as symmetrical? Because the count of eight fits the chemical facts, while six doesn't, and for no other reason. Lewis has described what happens, but he hasn't explained it. He is scornful of Bohr's approach, because it would require the electron to move in its orbit without generating an oscillating external field; Bohr himself, of course, was acutely aware of this problem. He claims, with no supporting arguments, that Bohr's explanation of the emission and absorption frequencies of hydrogen "may be translated directly into the terms of the present theory", with the relevant frequencies being "simply expressible in terms of ultimate rational units". In reality, there is no way in which an appeal to the ultimate rational units, while holding the electron in a fixed position, could have explained the factor of 4.0016 rather than exactly 4 in going from hydrogen to helium (see preceding chapter), but maybe Lewis was unaware of this refinement, or maybe what was good enough for Einstein wasn't good enough for Lewis. Or, more charitably, we may say that Lewis was more aware than any physicist of the complexities of chemical behaviour, hence his conviction that only the study of chemistry is capable of giving insight into the intimate structure of atoms. In either case, his impatient arrogance, whether on his own behalf or on behalf of his discipline, and his blanket dismissal of Bohr's entire line of reasoning, cut him off from his natural allies among the spectroscopists and physicists.

Now back to sodium chloride. It had long been suspected that bonding here involved transfer of one electron from sodium to chlorine, to give the Na^+ and Cl^- ions, as discussed above. Sodium is element number 11, and in Lewis's scheme it would have two electrons in the innermost part of the core, followed by a completed cube, and one spare electron. I shall write this as (2, 8, 1). Chlorine, on the other hand, is element number 17, with an arrangement (2, 8, 7). So after transfer of one electron, we are left with (2, 8) for Na^+, and (2, 8, 8) for Cl^-. Na^+ has the same electronic arrangement as the noble gas neon, element number 10, while Cl^- is *isoelectronic*, to use the modern term, with the next noble gas, element number 18, argon. Not surprisingly, magnesium, element 12, loses two electrons when it bonds to chlorine, giving the formula $MgCl_2$, with two positive charges on the magnesium cation balancing the total negative charge on the two chloride anions (overall, any compound has to be electrically neutral). In much the same way, oxygen (element number 8) has an arrangement (2, 6), and requires two electrons to become isoelectronic with neon. Thus sodium oxide is Na_2O, while magnesium oxide is just MgO and is built up of Mg^{2+} and O^{2-} ions. This counting scheme (actually, Lewis used more graphical schemes, involving drawings of cubes, and, later, arrangements of dots to represent electrons) makes sense of the existence of many ionic compounds, why they have the particular formulas they do, and the repetition of properties that gives rise to the Periodic Table. It is easy to understand, for example, why potassium with the arrangement (2, 8, 8, 1), and fluorine with the arrangement (2, 7), should so closely resemble sodium, (2, 8, 1), and chlorine, (2, 8, 7), respectively.

Two for the Price of One

But what about the bonding in sugar, whose solution in water does not conduct electricity? Or in organic compounds, such as methane? What is the nature of the bonding in water itself, for that matter? And what about the question that had bedevilled chemistry ever since Avogadro; what kind of force could possibly hold identical atoms together, as in the molecules of hydrogen, oxygen, or chlorine, H_2, O_2, and Cl_2 respectively?

This is where Lewis made his crucial suggestion. While atomic cores were (more or less) unchanged in chemical reactions, the outer shells were allowed to overlap. When this happened, the electrons in the region of overlap contributed to the cubic arrangement around *both* of the bonded atoms *at the same time*. So, for example, in the molecule Cl_2, we have a total of 14 outer electrons, but each of the electrons in the bond is double counted, giving each individual chlorine atom the energetically favourable total of 8 valence electrons around it. We have an *electron pair bond*, and have explained chemical combination in terms of the behaviour of the fundamental constituents of matter. To use our present-day terms, electron transfer gives *ionic bonding*, while electron sharing gives *covalent bonding*. Introducing a convention still in use, Lewis introduced diagrams in which outer shell electrons are represented by dots surrounding atoms, and shared electrons drawn in between them. Figure 12.1 shows these diagrams for the chlorine atom, the chloride ion Cl⁻, and a range of other important molecules.[15] Notice the double and triple bonds in the carbon dioxide and nitrogen molecules. These bonds are indeed much stronger than single bonds, hence the fact that burning carbon-containing fuel gives out energy, and the difficulty of separating the nitrogen atoms to make ammonia.

Bohr and others had already suggested the existence of atomic cores, and Bohr had suggested that in H_2 the electrons orbited around a point in between the nuclei, but these vague suggestions in no way detract from

:C̈l·	:C̈l:	:C̈l:C̈l:	H:H	H:C̈l:
Chlorine (atom)	**Chloride**	**Chlorine (molecule)**	**Hydrogen (molecule)**	**Hydrogen chloride**
Cl	Cl⁻	Cl_2	H_2	HCl

H H:Ö:	H H:N̈:H	H H:C̈:H H	:Ö::C::Ö:	:N:::N:
Water	**Ammonia**	**Methane**	**Carbon dioxide**	**Nitrogen (molecule)**
H_2O	NH_3	CH_4	CO_2	N_2

Figure 12.1.   Lewis diagrams for the chlorine atom, the chloride ion Cl⁻, molecular chlorine and hydrogen, hydrogen chloride, water, ammonia, methane, carbon dioxide, and molecular nitrogen. The atoms of hydrogen, carbon, nitrogen, oxygen, and chlorine contribute 1, 4, 5, 6, and 7 valence electrons respectively.

Lewis's claims to originality and priority, matters that were of great importance to him.

This scheme immediately explains why, in simple cases at least, compounds have the formulas they do. Sodium, magnesium, and aluminium combine with one, two, and three chlorines respectively because Na^+, Mg^{2+}, and Al^{3+} all have the stable (2, 8) arrangement of electrons, as in neon. Carbon, nitrogen, oxygen, and fluorine form the compounds CH_4 (methane), NH_3 (ammonia), H_2O (water) and HF (hydrogen fluoride) because in these compounds they achieve that same arrangement through electron sharing.

Lewis also discussed the reasons for the difference between what are called polar and non-polar compounds. Hydrocarbons such as gasoline, methane, or paraffin oil are non-polar, while compounds rich in bonds between hydrogen and oxygen, or hydrogen and nitrogen, are polar. Water is polar, while salts are the most polar compounds of all. Some compounds, such as aluminium chloride (see Chapter 9), can even show polar or non-polar behaviour, depending on their environment.

Lewis recognised that this was a difference of degree, rather than a sharp division between two different kinds. Different atoms have different tendencies to attract electrons towards them. This tendency is smallest in elements, such as sodium and magnesium, that have only one or two valence electrons, moderate for carbon, and very strong for atoms such as oxygen, fluorine, or chlorine, that have an almost complete octet. Hydrogen is just a little bit less electron attracting than carbon.[16] When the difference between the atoms in the bond is very large, we have virtually complete transfer, and, if the substance is dissolved in water, cation and anion go their separate ways. A moderately large difference, as in water, leads to an uneven distribution of charge, known as an electric dipole, and electric dipoles attract one another. Sugar is polar, because it has lots of O–H groupings bonded to its carbon backbone, so it dissolves in water, but does not dissociate or conduct electricity. Table salt dissociates in water, as already mentioned, to give Na^+ and Cl^- ions. Differences in polarity turn out to be of vital importance in understanding the chemistry of life.

Lewis's ideas were ridiculed at the time as illegitimate double counting. His chlorine molecule was compared to a husband and wife, each with six dollars in separate accounts, and two dollars in a joint account, and each claiming to be in possession of eight dollars. Organic chemists initially were not at all impressed. They had been writing formulas including bonds for decades. Back in 1874, Van't Hoff[17] had even worked out, from the number of isomers of different kinds, that the arrangement of four single bonds around carbon was tetrahedral. However, there was a strong prejudice that organic chemistry was somehow different, and its practitioners seem to have stopped worrying about what the bonds actually meant. Eventually, however, chemists interested in the detailed mechanism of chemical change took up the idea with great enthusiasm, using curly arrows to represent the way pairs of electrons moved around during reactions.

But How is this Possible?

But how does this sharing happen? And if it does happen, why just two electrons in the bond? Why not pile them all into the region in between the two atomic cores, where they would be attracted by both? Indeed, even more fundamentally, why don't all the electrons in any atom pile into the lowest energy inner shell, or orbital, since this is the one of greatest stability? The answers were to emerge from the quantum theory, of which Lewis had been so dismissive, over the next dozen years.

Firstly, the nature of sharing. This comes out of the application of wave mechanics to a system containing two atoms, with orbitals that overlap. It can be shown that when this happens, the individual orbitals cease to exist and we are left with two combined orbitals spread over both atoms. One of these is bonding (meaning that it is of lower energy than the original separate orbitals), and one anti-bonding, at higher energy.[18] The electrons go into the bonding orbital, and this is what holds the atoms together. Anti-bonding orbitals can be ignored for the moment, as far as we're concerned, but will turn out to be very important later when we discuss where colour comes from.

But even knowing all this, we still need to ask, why just two electrons in an orbital, what stops all the electrons from piling into the lowest energy orbital, and how many orbitals are there anyway?

The number of different possible orbitals in an atom comes out of the mathematics of the number of different ways you can wrap a wave function round a central atom. It turns out that there are only so many different ways of doing this so that it joins up properly; these are called "spherical harmonics" and would match up to the possible vibrations of a spherical drum, if we could make such a thing.

The lowest set of orbitals has got just one member. The next set has four (1 + 3) members. The third set again gives us 1 + 3 members of roughly the same energy, and another set of five members as a kind of mezzanine, reaching up to roughly the same level as the beginning of the fourth set.[19]

The remaining questions were answered by Wolfgang Pauli, an Austrian of Jewish descent, and a member of the group of physicists centred around Bohr. He spent most of his career at ETH Zurich, with a break (including the period of World War II) at the Institute for Advanced Study in Princeton. He comes across as among the most intriguing of Bohr's circle, at once anti-metaphysical and mystical, enormously likeable and yet given to using words as sharp instruments, one of the leading physicists of his generation, and yet someone that no colleague would allow anywhere near his measuring instruments.

Pauli's mere presence was said to be enough to upset an experiment. On one occasion, according to legend, apparatus in the laboratory at the University of Göttingen failed to work for no apparent reason. *It was subsequently discovered* that Pauli had been passing through Göttingen by train at the time. Pauli enjoyed such stories as examples of synchronicity, a term coined by the psychoanalyst Carl Jung for the situation where events are causally unconnected, but nonetheless occur together in a psychologically meaningful manner. Pauli was himself analysed by one of Jung's associates, corresponded with Jung for many years, and sent him descriptions of over 1000 dreams. More interestingly, perhaps, they discussed the formation of scientific concepts, and how these were related to psychological archetypes. For Pauli, this was not some mere peripheral activity or pastime. In his opinion, "there must be very deep connections between soul and matter and, hence, between the physics and the psychology of the future ... because otherwise the human mind would not be able to discover concepts which fit nature at all."[20] And again, "it is my personal opinion that in the science of the future reality will neither be 'psychic' nor 'physical' but somehow both and somehow neither."[21] Pauli also coined two of the harshest criticisms ever uttered by one scientist of another. Of a junior colleague with whom he was not impressed, he said "So young, and already so unknown." The name of the colleague has not come down to us. And of a theory totally devoid of merit, even the merit of being experimentally testable, he said "Not even wrong," an expression that retains its usefulness.

When trying to make sense of the detailed structure of atomic spectra in the presence of a magnetic field, Pauli came up with the idea that an electron had a magnetic moment, which could only be "up" or "down" with respect to a measuring field, or to the magnetic field of another electron. This property is often referred to as "spin", because in classical physics a rotating electrical charge would indeed act as a magnet, although such mechanical analogies are not really applicable. One surprising but enormously important consequence of this turned out to be, that any one orbital had room for just two electrons, with spins pointing in opposite directions. Pauli compared them with seats on a bus.[22] The lowest level, with just one orbital, is completely filled in element number 2, helium. The next level, with four orbitals, is filled in going from lithium, (2,1), to neon, (2, 8), and the four orbitals after that in going from sodium (2, 8, 1), to argon (2, 8, 8). The next two elements, potassium, K, (2, 8, 8, 1) and calcium, Ca, (2, 8, 8, 2), behave exactly as expected, losing either one or two electrons in their chemical reactions to give K^+ and Ca^{2+}, both isoelectronic with argon. (Why K for potassium? Thank the Arabs. Al-kali, the-pot-ash. Wood ash is rich in potassium carbonate, giving us Kalium as the German name for potassium.) After that, what I have called the mezzanine of the third level presents itself, with five orbitals and therefore room for 10 electrons, and things get more complicated.

On the strength of Einstein's nomination, Pauli received the Nobel prize for physics in 1945 "for the discovery of the Exclusion Principle, also called the Pauli Principle", which, together with the closely related concept of electron spin, bridges the gap between the study of individual particles, and atomic and molecular structure.

"Science has its cathedrals," wrote Lewis in the introduction to his textbook on thermodynamics. Within the stone cathedral, we are inspired when different lines of masonry, starting out from foundations far apart, meet at a common point in a majestic Gothic vault. It is equally inspiring when different lines of scientific evidence and reasoning, starting out as far apart and in such different style as the empirical generalisations of chemistry, and the mathematical theories of physics, meet, as here, in a common conclusion, a shared understanding of the nature of chemical bonding.

More about Lewis

During World War I, which the US did not enter until 1917, Lewis enlisted and was sent to Paris to work with the United States Chemical Weapons Service. There he organised the training of gas officers, something that had previously been shockingly neglected, and American gas casualties dropped dramatically. In World War II, however, he was left on the sidelines while his colleagues and junior faculty were deeply involved in war work, and especially in the Manhattan Project that was to lead to the development of the first nuclear weapons. This was in part because of his personality (Oppenheimer could not see him as a potential team player), and may have contributed to a growing depression.[23]

Experts tell us (do we really need experts to tell us?) that in the end, getting what we want doesn't make us any happier. However, not getting what we think we deserve can certainly make us more miserable. One eminent chemist who thought he deserved the Nobel Prize used to embarrass everyone by planning his departmental party for announcement day. Another, with stronger claims, gave up chemistry for novel writing. But none could have felt so much passed over, and with such good reason, as Gilbert Lewis. If you ever studied chemistry, you have probably heard of Lewis structures, and Lewis acids. If you have studied thermodynamics, you will have used Lewis's formulation of the basic equations, and his concept of *fugacity*. He was also responsible for the distinction we make between fluorescence and phosphorescence; phosphorescence is longer lasting because it involves a change in the spin (see above) of an electron. Lewis greatly minded not getting the Nobel Prize, and with hindsight we have to work quite hard to understand why he didn't. The Prize is not a lifetime achievement award, but is given for one specific advance or discovery. There are in fact deep connections between Lewis's many different chemical achievements, but this was perhaps not obvious to lesser minds at the time. It was left to Lewis's contemporaries and juniors, such as Irving Langmuir and Linus Pauling, to spell out the full implications of his ideas on bonding, and his theory was (on occasion still is) sometimes incorrectly referred to as the "Langmuir–Lewis theory". Personalities may also have been involved.[24] Such things would matter to anyone, but it seems that they mattered very much to Lewis; we have seen how long he bore a grudge against Richards.

Irving Langmuir was awarded the Nobel Prize in chemistry in 1932 for his work on surface oil films, which under the right circumstances form a layer just one molecule thick. On 23rd March, 1946, he received an honorary degree at Berkeley. Lewis, for whatever reason, did not attend the ceremony, but Langmuir and Lewis had lunch together afterwards. Later, Lewis was found dead in his laboratory, where he had been working alone. The laboratory was full of hydrogen cyanide fumes, and there was a broken ampoule of the substance next to Lewis, but Lewis had been planning to work with this substance in any case, there was nothing unusual at that time about him handling it in the open laboratory (nowadays, no one would dream of doing so outside a fume hood), and we will never know whether we are looking at a simple laboratory accident, a stroke or heart attack that led Lewis to drop the ampoule, or suicide.

Shape

I promised to tell you why things have shapes, and it is high time for me to do so. Very big things, like stars and planets, have roughly spherical shapes, because of gravity. Everyday things have the shapes they do because of the way that their chemical constituents interlock. This in turn depends on the forces between molecules, and the forces inside molecules. The forces between molecules depend on whether the individual molecules contain polar bonds, and, if they do, the attraction between the more positively charged end of one polar bond, and the more negatively charged end of another. Within a piece of solid metal, the shape will depend on a promiscuous sharing of valence electrons over many atoms. The particular arrangement adopted will depend on the number of valence electrons per atom, and whether they all come from the same energy level, in ways that are quite well understood. Within a crystal, such as sodium chloride, the shape depends on the tendency for each cation to surround itself with as many anions as possible, and vice versa, while cation repels cation and anion repels anion. So simple salts tend to adopt very simple shapes. The details depend on the relative sizes of cation and anion (for example, if the cation is very much smaller than the anion, it only has room for four near neighbours, while if they are of the same size, there is room for eight). At least, that is what I was taught. It now turns out that overlap between the outermost electrons of the anion, and the empty orbitals of the cation, also play a role.

And finally to the most important case, shape as dictated by the arrangement of the electron pairs around an atom. Sticking to the cases where the octet rule applies, we have four pairs of electrons, and since these are negatively charged they will repel each other. The geometrical arrangement that will keep them as far away from each other as possible is a tetrahedron. So the hydrogen–oxygen–hydrogen arrangement in a water molecule will be bent, with oxygen roughly at the centre of a tetra-hedron, two corners occupied by hydrogen atoms, and the other two by the *lone pairs* of electrons, the two pairs not involved in bonding, as shown in Figure 12.1. Ammonia will resemble a pyramid with three corners occupied by hydrogen atoms, while the lone pair on nitrogen points towards the fourth. Methane will have a tetrahedral shape, explaining what the organic chemists had long figured out from purely structural arguments.

To Sum Up

The work of Henry Moseley gave experimental underpinning to the con-cept of atomic number. Gilbert Lewis used this concept, and his own curi-ous model of the atom, to formulate the ideas of what we now call ionic and covalent bonding. Lewis's thinking was based on the idea that an octet of electrons conferred special stability, and subsequent develop-ments in quantum theory, in particular the development of wave mechan-ics and the *Pauli exclusion principle*, explained why this should be so.

Chemical observation had led to the observation of regularities in chemical behaviour when atoms were arranged (more or less) in order of increasing weight. Moseley's work proved that the more fundamental quantity was atomic number, equal to the charge on the nucleus, and hence to the number of electrons around it. Lewis suggested that these electrons were in general arranged in groups of eight, and that this octet had particular stability (although the innermost group, filled in the noble gas helium, contains only two). In Lewis's scheme, there are two extreme kinds of chemical bonding, ionic and covalent. In an ionic bond, electrons are transferred from one atom to another, thus (in many cases) leaving both with completed octets. A covalent bond involves sharing of electrons, and these shared electrons count towards the octets of both of the atoms involved. The simplest kind of covalent bond, a single bond, involves just two shared electrons. In non-polar compounds, the bonds are between atoms that have similar abilities to attract electrons. In polar compounds, the electrons involved in bonding are attracted more strongly to one atom than the other.

Subsequently, wave mechanics explained the special role of the numbers 2, 8, and higher numbers that become important for the heavier elements, in terms of the ways that the wave function of an electron can fit round a nucleus. This gives us the number of orbitals available at each energy level. Sharing is explained by mathematically combining wave functions on adjacent atoms, to give an orbital spread out over both of them. Finally, the number of orbitals used is determined by the number of electrons to be accommodated, and the fact that each orbital can hold just two electrons (one for each of the two possible directions of spin).

Shape depends on how electrons are shared, and the fact that electron pairs repel each other. Specifically, when carbon is bonded to four different atoms, the bonds point towards the corners of a tetrahedron, a fact of central importance in the chemistry of life.

13 Why Grass is Green, or Why Our Blood is Red

What hope have we to know ourselves, when we
Know not the least things, which for our use be.
Why grass is green, or why our blood is red,
Are mysteries which none have reach'd unto.

John Donne,
"On the progress of the soul;
the Second Anniversary", 1612

John Donne, admirable poet but deplorable philosopher, subscribed to the doc-
trines of Intelligent Design, and faith through ignorance. Can we do better, as
we approach the four hundred and second anniversary?
We can answer Donne's question at levels that he himself could not even have
imagined. Why do things absorb light? Why do green leaves absorb the par-
ticular light they do, and what do they do with it once they've got it? How
come this light is available to them? Why does this look green to our eyes? How
would it look to John Dalton's eyes, or to the eyes of a chimp, a monkey, a bee,
a deer? And finally, a question that happened to be the subject of one of my
own first research papers,[1] why is the colour of blood so different?

The Nature of Colour

Let's take these questions in order. The reason why things absorb visible
or ultraviolet light is just an elaboration on the one that Niels Bohr gave

to explain the line spectrum of the hydrogen atom. During absorption, an electron jumps from one level to a higher one, with the separation between them equal to the energy of the photon absorbed. The absorption spectra of single atoms consist of nothing but sharp lines, but for molecules, these lines are spread out, because the electron jump usually weakens bonds in the molecule, changes the equilibrium distance between its atoms, and sets it vibrating over a whole range of different frequencies.

To understand where these higher energy levels come from in a molecule, we need to go back to the simple theory of the electron pair bond. Consider H_2, the simplest possible molecule.[2] Lewis, you may recall, was ridiculed for suggesting that the electrons in the bond could be double counted, as belonging at one and the same time to both the hydrogen atoms. Quantum mechanics, however, came to the rescue here. Take the wave functions for the lowest energy orbitals of the two hydrogen atoms. The rules of quantum mechanics allow you to combine these together, either by adding them (in phase), or subtracting one from the other (out of phase). A few lines of elegant algebra are then enough to show that the "in phase" combination is at lower energy than the individual atomic levels (bonding), while the "out of phase" combination is at higher energy (antibonding).[3] So the hydrogen molecule can absorb light, making an electron jump from the bonding to the antibonding level; because of the large energy gap between them, this light is in the ultraviolet. When this happens, antibonding cancels out bonding, there is no longer anything holding the molecule together, and the two hydrogen atoms go their separate ways. This is an example of *photodissociation*, which we met before when talking about how ultraviolet light breaks up oxygen molecules in the stratosphere, on the way to making ozone.

Things get more interesting when you join up more atoms, especially if you can make rings of atoms with alternating single and double bonds. Now, when you do the algebra, you find that there are lots of different possible energy levels, with the electrons spread out in different ways all round the ring. The important thing for us is that the energy gap between the highest occupied energy level, and the lowest unoccupied, is a lot less than it is in smaller systems. So instead of the molecule only absorbing in the ultraviolet, it absorbs light over one or more frequency ranges in the visible. The light we see is the light that *isn't* absorbed, but this is only part of the full range that makes up white light, and so the molecule is coloured. The exact colour will depend on the details of the atoms in the ring, and the other groups of atoms bound to them, but this, basically, is what is responsible for the colours of natural and synthetic dyestuffs of all kinds.

So now we know why grass is coloured; because at the heart of the chlorophyll molecule there is a ring of alternating single and double bonds. Chlorophyll — from Greek chloros, green, and phyllon, leaf.

Sunlight to Sugar

Why green, specifically? Because chlorophyll absorbs at the blue end of the spectrum, and also at the red, leaving the green to be scattered off a leaf and strike our eyes.

Actually, this is a pretty unsatisfactory arrangement. The chlorophyll molecule is failing to absorb in the very region where the sunlight is most intense. However, living things are the products of evolution, not design, and must use the molecular structures, such as the particular alternating single and double bond arrangement of the *porphyrin* ring at the heart of chlorophyll, that are easiest for them to make. True, the overall efficiency of the system is to some extent boosted by "antenna molecules", such as carotene, which absorb blue light and channel the energy to the photo-chemical reaction centres. Nonetheless, much of the light energy is simply wasted. Plants do have available to them ways of making pigments to cover the entire visible spectrum (think of a flower garden), but the coupling of these to photosynthesis just does not seem to have happened. The overall efficiency of photosynthesis is only around 2%, as opposed to the 27% obtained in some artificial solar energy generators,[4] and a really efficient plant would look black.

The absorbed light drives photosynthesis, a complex process with many reaction steps. The overall result in plants is to convert water and carbon dioxide to oxygen and organic fuel, such as glucose, which is either used to drive the reactions of life, or gets converted into other materials. All green plants carry out this process. So do one family of bacteria, the cyanobacteria (cyan, from Greek kuanos, dark blue; as in cyanide, so named because hydrogen cyanide was originally isolated from Prussian Blue), using the same molecular machinery.

There are five separate steps to this process. In the first, light is absorbed by a bunch of chlorophyll molecules in a structure known as

Photosystem II or PSII. The energy of the absorbed light raises electrons to a high energy level, spread out over these molecules.

The high-energy electrons are then transferred through a chain of intermediates, being made to do useful work on the way, to a second structure, containing slightly different forms of chlorophyll, known as Photosystem I or PSI (my apologies for these confusing names).

This also absorbs light, of slightly different wavelength, and the electrons, now at higher energy than ever, are used together with water to convert organic molecules to a hydrogen-rich form. This hydrogen is eventually transferred through intermediates to molecules that take up carbon dioxide, eventually converting it to glucose.

Meantime, the original PSII structure picks up electrons from what is known as the "oxygen evolving complex", a cluster of four manganese atoms. Manganese, like other metallic elements in the middle of the Periodic Table, cannot[5] gain or lose enough electrons to match the number of a noble gas, as discussed in Chapter 12, but instead can form a range of different kinds of cation, by losing different numbers of electrons to its surroundings, depending on circumstances. With four manganese atoms, each initially carrying a charge of +2, and each having given up one electron to the chlorophyll in PSII, we have just the right amount of excess positive charge to convert water to oxygen.[6]

Putting it all Together

So how could a system requiring so many separate parts come into being? The answer, as with all such complex systems, is one part at a time.[7] Oxygen-producing photosynthesis uses water as the ultimate source of the electrons that are eventually used to convert carbon dioxide to glucose, but there are simpler, and almost certainly older, forms of photosynthesis that use hydrogen sulphide, or electron-rich compounds of iron, instead. There are whole families of photosynthetic bacteria using these, and containing only PSI or PSII.[8] Moreover, we know from genome *sequencing* (essentially, the same kind of DNA evidence that is used every day in our courts to establish paternity) that PSI and PSII share a common ancestry, and may have evolved along separate pathways in bacteria living in different environments. If so, they came together in some ancestral cyanobacterium by the familiar process of horizontal gene transfer.[9] This is a well studied process, responsible among other things for the spread of antibiotic resistance between different pathogens. The oxygen-evolving manganese oxide system is reminiscent of manganese-containing minerals, and the earliest version could have formed spontaneously inside the cell by the operation of simple inorganic chemistry.

Finally, on this topic, how did PSI and PSII find their way into green plants? By the process known as *endosymbiosis* (Greek: endo, inside; sym, together; biosis, way of living). Their chlorophyll-containing *chloroplasts* are the descendants of cyanobacteria that invaded, established their homes in, and threw in their lot with more complex cells to form the green algae that were the ancestors of all plants, some time around a billion years ago. A 2011 analysis of the cell contents specific to photosynthetic organisms confirms this explanation.[10]

Where does the light energy come from? From the Sun, of course, and from nuclear fusion reactions deep within it. Sunlight covers the range it does because the surface of the Sun is at the temperature that it is. Stars larger than our sun give out energy faster, at higher temperatures, so their light is shifted towards the blue. Smaller stars show the opposite effect, and are red dwarfs. Does this make the Sun particularly favourable for the development of life on its planets? Not really. A smaller star would have its maximum output in the red, where it can be more efficiently absorbed by the readily synthesised porphyrin-type molecules. It would also have a far greater lifetime, during which the emergence of life and its evolution of complexity could occur.

The Bird's Eye View is Different

Of course, when we say grass is green, that is not just a statement about grass, but a statement about how our eyes work. We cannot see into the infrared, for the excellent reason that infrared photons do not have enough energy to cause the required chemical changes in the light-sensitive cells of our retinas. Large animals like us cannot see into the ultraviolet, because ultraviolet light is absorbed by the eye before it reaches the retina. Humans, most of us, like our close relatives the apes (pedants might insist that we *are* apes, and they would be right),[11] and our rather more distant relatives, the Old World monkeys, have three-colour vision. There are three primary colours of light as far as we are concerned, and these are, perhaps surprisingly, blue, green, and red, corresponding to the different properties of the light-detecting pigments, or opsins, in different cells of the retina. This is rather different from the familiar blue-yellow-red scheme that we are familiar with, but this is because when we are mixing pigments, we are subtracting light, not adding it. However, I found it impossible to believe that green light plus red light looks yellow, until I saw it demonstrated using a colour projector. Unlike a chord in music, our visual perception is a more complex construction from the stimuli than a mere addition. The retina is not just a collection of photoreceptors; it is an outcrop of the brain, with connections between the different kinds of light-sensitive cell, and the information that it sends to the brain proper along the optic nerve is already highly processed.

Insects have far smaller eyes than we do, except in old-fashioned horror movies, which means that they can detect ultraviolet light.[12] Bees have three opsins, sensitive to UV, blue, and green. Birds and reptiles have the correct genes to synthesise at least four opsin genes, but mammals in general only have two, suggesting loss of function during the long period when dinosaurs ruled the Earth, and mammals were generally nocturnal. However, Old World monkeys, and apes, including humans, have, as mentioned, three colour vision. The red- and green-sensitive opsins are very similar to each other, differing in only a small number of the amino acids in the protein that cradles the light absorbing part of the molecule, and determines where it absorbs. Something similar has happened independently to some New World monkeys. The inference is that the red-sensitive opsin arose through a gene doubling event, followed by selection for division of labour between the two copies. This is a very common mechanism by which evolution generates new capabilities. Duplicating chunks of DNA is a common kind of error. When a useful gene is duplicated, selection pressures keep one of them the same, leaving the other free to explore other possibilities, and to be selected for new capabilities. Animals with three colour vision can select red over green, tender leaves over tough, ripe over unripe. So to go back to our original questions, the world looks very much the same to a chimp or a New World monkey as it does to us, but to a deer or to John Dalton, who was red-green colourblind, it is less visually interesting. One implication is that red camouflage would be as effective as green camouflage for deer hunters, and greatly improve their safety, but I don't expect the idea to catch on. As to how the world looks to those creatures with four or even more opsins, and therefore four or even more primary colours, I cannot begin to imagine.

And Now for Donne's Other Mystery, Why Blood is Red

The porphyrin molecular motif, central to chlorophyll, is found in all living things on Earth, in molecules that regulate the transfer of electrons. The innermost part of the motif consists of four nitrogen atoms, each with a lone pair of electrons pointing towards the centre. These electrons can attract and hold a positively charged metal ion, and in chlorophyll itself that ion is doubly charged magnesium, Mg^{2+}, with the same number of electrons as the noble gas neon. In this ion, all the electrons are tightly held, there are no low energy empty orbitals, and so the magnesium contributes nothing to the colour.

Distant relatives of chlorophyll, in evolutionary terms, are those porphyrins that contain doubly charged iron instead of magnesium. Iron, like the manganese that we met earlier in this chapter, is one of the elements in the middle of the periodic table, and can fairly readily lose another electron to become triply charged. The electron transfer properties of iron porphyrin–protein assemblages are fine tuned by the interactions between the porphyrins themselves, and the proteins in which they are embedded. Electron transfer between doubly and triply charged iron in such systems plays an important part in photosynthesis as in many other electron cascade processes.

In haemoglobin, the molecule that gives blood its colour, the situation is subtly different. Haemoglobin also consists of an iron-containing porphyrin embedded in a specific protein. However, its function is to transport molecules of oxygen around the body, rather than to take part directly in electron transfer processes. Throughout, the iron remains in the doubly charged state, which interacts strongly enough with electrons on the oxygen to keep it in place, but not so strongly as to prevent its release where required. The colour of haemoglobin, and of all iron-containing porphyrins, comes from two sources; electron jumps within the porphyrin ring, now modified by the interaction between the ring-based orbitals and the loosely held electrons on iron, and "charge transfer" absorption bands, where one of these electrons is made to jump to a higher energy empty orbital on the ring. The overall effect is to move the position of the absorption band in chlorophyll away from the red, so red light is now what we see.

And what lipstick manufacturers take advantage of, because full red lips are advertisements of good health in a prospective mate, and have therefore been selected in humans (much more than in other apes, oddly enough) as sexual adornments. John Donne himself greatly appreciated the joys of sex, and I like to think that he would have enjoyed this answer to the questions that he posed.

To Sum Up

Things have the colours they do because they absorb light, and the colour we see is the colour of the light that is not absorbed. Grass is green because it contains chlorophyll, and chlorophyll is the molecule that absorbs light from the Sun in the process of photosynthesis, on which almost[13] all life on Earth depends. Green is one of the primary colours in our three-colour vision, which we share with Old World monkeys and apes, but insects, birds, and reptiles have slightly different and in many cases richer systems. Blood contains haemoglobin, which carries the same porphyrin motif as chlorophyll, but its light-absorbing properties are modified by a doubly charged iron ion at its core, leading to the difference in colour.

John Donne showed insight in his choice of questions, but gave hostages to fortune when he let so much hang on our inability to answer them.

14 Why Water is Weird

Liquid water is one of the most common, most important, and most strangely behaved substances on the surface of this planet.

Going Liquid

Liquids are always more difficult to understand than solids or gases. In a gas, we have small, weakly interacting, molecules moving at random, and bouncing off each other. Small compared with the space between them, and with their mutual influence small compared with energy of motion. So we can get away with a very simple mechanical model, the "ideal gas", that ignores these small effects altogether. This model explains the mathematical laws (Boyle's Law, Charles's law, Avogadro's hypothesis) that, to a good approximation, describe the relationships between pressure, temperature, density, and molecular weight. All real gases deviate from this idealised model, but we can use it as a starting point and then feed in the reality of molecules of finite size, weakly interacting with each other, to explain the deviations from idealty. In solids, we have the opposite situation. The units that make up the solid are held firmly in their particular individual places. Their thermal energy sets them vibrating around the most stable positioning, but is not enough to allow them to move past each other. So solids adopt definite structures, determined by the shapes of their constituent units, and how they can most efficiently pack together. Liquids are somewhere in between. Their components can slide over each other, but cannot escape each other's influence completely. The way they can move over each other depends on the shape of the molecules, and the polarity (amount of unequal electron sharing; see Chapter 12) of the bonds within them. The way that the molecules orientate themselves with respect to each other is anything but random. Often, there are regions within the liquid where several molecules have the most stable mutual orientation, much as in the solid, and such regions are continually re-forming and collapsing.

The Hydrogen Bond

Now think of a molecule of water, H_2O. Using Lewis's theory, we can see that there will be just four pairs of electrons around the oxygen, two of them in the electron pair bonds between H and O, and two lone pairs (as in Figure 12.1), and these four will point roughly towards the corners of a tetrahedron. So the water molecule is bent. In addition, each of the O–H bonds is polar, with the electron pair pulled more towards oxygen than hydrogen, because oxygen is more electronegative,[1] leaving some positive charge on the hydrogen, while the negative charge on oxygen is concentrated in its lone pairs.

As the result of all this, there will be a strong electrical attraction between the molecules in liquid water, and a tendency for them to orient themselves towards each other in a particular way. The most stable arrangement will be one where the hydrogen atom of one water molecule is close to, and indeed overlaps, the region of space occupied by the oxygen lone pair on another molecule. This gives rise to what is called a *hydrogen bond* between the two molecules. Such "bonds" are not as strong as typical chemical bonds, but are nonetheless strong enough to more or less hold molecules in place. Hydrogen bonding occurs quite generally between hydrogen atoms directly bonded to nitrogen, oxygen, or fluorine, and the concentrated lone pairs found on such atoms.

Why Ice Floats, and Why it Matters

We can now begin to make sense of the odd properties of water, starting with one of the oddest, why ice floats.

When something melts, the liquid usually takes up more space than the original solid. This is exactly what we should expect. The atoms and molecules in solids pack regularly, with efficient use of space, while in liquids the separate units can move past each other, leaving loose ends and empty spaces. But in ice, each molecule of H_2O is hydrogen bonded to a total of four others, two by way of its hydrogen atoms to oxygen lone pairs, and the other two, by way of its own lone pairs, to the hydrogen atoms of neighbouring molecules. The result is to give an open frame-work structure. Melting leads to a partial breakdown of the hydrogen bonded framework, allowing closer packing of the molecules. So ice floats. In fact, it is about 10% less dense than liquid water at 0°C, which is why the top 10% of an iceberg shows up above the surface. Just above the melting point, however, some well ordered local regions still persist, although these progressively disintegrate as the temperature is raised. The result is that for about 4° above its melting point, water actually con-tracts as it is heated.

All this is good news for fish, and bad news for Russians.

Consider what happens in a pond as winter approaches. If water behaved like most substances, cool water on the surface would sink until the entire pond was at the freezing point, and would then freeze solid from the bottom up. As it is, cool water sinks until the entire pond is around a temperature of around 4°C, at which point surface water starts expanding as it cools further, and finally deposits a layer of ice, a poor conductor of heat, which acts as a protective blanket for the liquid water underneath.

Now look at a map. Floating ice covers the Arctic, frustrating dreams of a North West Passage from the North Atlantic to Asia round the top of Canada,[2] and sealing up Russia's long northern coastline for several months each year. Ice also blocks the Baltic Sea each winter, making the port of St. Petersburg useless during that time. This explains the importance to Russian policy of a warm water port, driving expansion to the Black Sea by 1794, war with Britain, France, and Turkey in 1854,[3] a continual jostling for influence in the Middle East throughout the 19th and 20th centuries, and, most recently, tension between Russia and the newly created state of Ukraine over the Crimea and the rights of the Russian fleet in Odessa.

Floating ice has a huge impact on the Earth's climate. Ice increases the planet's *albedo*, or ability to reflect energy back to space. The more ice there is, the more sunlight Earth reflects, cooling the planet down and making more ice; and vice versa. This is a potential source of instability (technically, a *positive feedback loop*), which helped drive the advance and retreat of the great Ice Age glaciers.

Ice melting can for this reason be expected to increase the impact of any greenhouse effect due to human activity. Global warming has caused glaciers and the polar ice sheets to shrink, exposing more of the land and open water that lies beneath them. These absorb more of the energy in sunlight, thus causing yet more warming. The shrinking of Alpine glaciers over the past fifty years, and the recent collapse of large parts of the Antarctic ice sheet, lead climatologists to conclude that this process is already well under way.

The low density of ice also contributes to the pleasure of skaters. The ice directly underneath the skater's blade is under a lot of pressure. This pressure shifts the melting point of ice towards lower temperatures,[4] which will tend to cause surface melting directly underneath the blade. So the skater's path is lubricated by a thin film of liquid water. Actually, there is more to it than that. Scholars differ on the question of just how much melting takes place under these conditions, although skaters' tracks are sometimes clearly visible on the rink. More important may be the fact that the atoms on the surface layer of the ice have a shortage of neighbours to which they can hydrogen bond, so that this outermost layer, at temperatures just below freezing, is incompletely bound and more like a liquid than a solid.

Other Weird Liquids

There are a few other materials that expand on cooling, and these include silicon and germanium, antimony, and silica (silicon dioxide). Silicon and germanium have the same arrangement of outer electrons as carbon, which means that each atom can make four single bonds. So in the solid, once again, we have an open arrangement, with each atom having four nearest neighbours, much as in diamond.[5] The bonding electrons are more loosely held than in diamond, so that diamond is an insulator, while germanium and silicon are semiconductors. Indeed, the semiconductor properties of silicon are fundamental to nearly all our everyday electronic devices, with germanium being used for special applications. When molten, however, the atoms pack much more closely, hence the increase in density, with each atom having up to twelve neighbours. There are more possible "bonds" than there are electrons to fill them, so that the electrons can move readily from one position to another, and these liquids are metallic conductors. Solid antimony has five outer electrons, meaning that it can readily form three bonds to its neighbours, and satisfy Lewis's octet rule. This it does in the solid by arranging the atoms in sheets, each with three nearest neighbours. This again is a space-consuming arrangement, so antimony is also denser as a liquid than as a solid. Antimony and its alloys also have relatively low melting points, a combination of properties that made them useful for preparing sharp-edged printers' font, as the solidifying metal expanded in its mould.

In water, each oxygen makes hydrogen bonds to a total of four oxygen neighbours, two through its own hydrogen atoms, and two through its lone pairs. In silica, SiO_2, each silicon is connected to four silicon neighbours by oxygen atoms, each oxygen forming a bridge by making two separate single bonds. So once again the melt is of higher density, adopting a more compact structure through breaking bridges. Once again, we have an open ideal structure because of a tetrahedrally bound atom, and once again the less than perfect melt is denser than the solid.

This leads to an important difference in behaviour between the two main types of igneous rock, basalt and granite. Basalt is formed from "basic" melts, where there are enough metal oxides to combine with oxygen, and the silicon present, to make metal silicates. These minerals, like most materials, are less dense when liquid. As a result, a reduction in pressure leads to a lowering of their melting points, since, by Le Châtelier's rule, lower pressure should favour the less dense form. So basalt flows more freely on the surface than when under pressure at depth, giving rise to the flowing streams of lava in volcanic eruptions and in mid-ocean vents, and to the typical lava plateaus and volcanic cones of basalt landscapes. Molten granite, on the other hand, is more "acidic", meaning that it contains free silica, so that when the pressure is reduced, it becomes stickier as the silica starts solidifying. So the lava in silica-rich, granite-producing volcanoes tends to set firm as it rises, forming irregular masses in magma chambers below the surface. These, when exposed by weathering, give rise to dramatically rugged landscapes, as in Big Bend National Park in Texas, the crags of the Sandia Mountains, or (nearer to my present home) Ailsa Craig, the granite plug of an ancient volcanic vent, resembling nothing so much as a giant upside down pudding basin.

Wet and Wonderful

Next, why water wets so many things, and dissolves so many others. This again is a matter of polarity, and, very often, of hydrogen bonding. Take the familiar example of table salt, NaCl. When this dissolves, the positive charge of the Na^+ ion is attracted to lone pairs on oxygen, while the negative charge of Cl^- is attracted to the positive charge on hydrogen; in fact, there is hydrogen bonding between the water and the Cl^-. Sugars have a lot of O-H groups attached to a carbon backbone, and these also take part in hydrogen bonding to water when they dissolve. The cellulose in wood and cotton is built up from sugar units, and inherits many of their O-H groups, so water wets them. Rocks, generally speaking, are wettable, because they contain metal cations that attract the oxygen end of water molecules, and silicate or carbonate anions that hydrogen bond to water through their own oxygen atoms. Many organic plastics would ideally not be wettable by water, but even in cases like these, surface oxidation puts polar groups on the surface of the plastic, which is why milk wets its polyethylene container.

Next, the high surface tension, high specific heat, and long liquid range of water. All of these are consequences of hydrogen bonding. Water molecules attract each other strongly, but those on the surface have less than their full complement of neighbours. Infrared spectroscopy (which measures how tightly atoms are held in place) shows that for the molecules on the surface of liquid water, one hydrogen is bonded to the bulk water directly beneath the surface, while the other protrudes above it.[6] So these surface molecules are less stabilised by hydrogen bonding, making the surface a region of higher energy, and leading to a strong tendency to make that region as small as possible, and hence to a high surface tension. Surface tension makes wet things cling to each other, and pulls water into little drops on greasy surfaces. It also makes it possible for water beetles to run across ponds without breaking the surface. Water has a high specific heat, meaning that it takes a lot of heat energy to change its temperature. You can heat oil on top of a stove to frying temperature (200°C) more quickly that you can heat water to boiling (100°C). This is because raising the temperature breaks more of the hydrogen bonds. To evaporate water, you need to break all the hydrogen bonds between the H_2O molecules, and this requires even more energy. That is why the boiling point of water is very high for such a small molecule (for comparison, the boiling point of methane, CH_4, is $-161°C$), and why sweating is such an effective cooling mechanism. Of course, sweating is a lot less efficient when the air already holds almost as much water as it can at that temperature, which is why 30°C (86°F) in the humidity of Washington DC is so unpleasant, while 30°C in Spain or New Mexico is quite enjoyable.

Next, the many forms of water. The structure of water depends on the temperature (higher temperatures tend to favour the less tightly structured forms, even although they have less efficient hydrogen bonding), and the pressure (higher pressures favour more densely packed arrangements). The result is that in addition to ordinary ice, liquid water, and water vapour, there are at least 9 other forms.

Fact, Fraud, and Fiction

Some non-existent forms are worth mentioning. A few years ago, some Russian scientists claimed to have made a more stable form of water, known as polywater, with a higher boiling point than ordinary water. This if true would have been the most alarming of news. For polywater could act as a template for normal water to rearrange itself, totally altering its properties, on which so many things (including life itself, as we shall see) depend. This nonsense got far more attention than it deserved; after all, if there really were a more stable form of water, surely it would have come into existence naturally some time in the more than four billion years since the formation of our planet? However, it was only finally laid to rest, when at age 90 the veteran electrochemist Joel Hildebrand showed that the elevated boiling point of "polywater" was due to the presence of large amounts of dissolved impurities. The idea of a more stable form of water had already occurred, as a device to be used in fiction, to Irving Langmuir, G. N. Lewis's contemporary. Langmuir mentioned it to H. G. Wells, who didn't use it. Much later, Langmuir's former student, the atmospheric scientist Bernard Vonnegut, passed on the idea to his brother Kurt, who used "ice-9" as the central plot device in his catastrophe novel *Cat's Cradle*.

Other purely fictional forms of water remain on sale to a gullible public. We have "magnetised" water, that has been passed through a strong magnetic field, and is supposed to be healthier and less corrosive. There are in fact some possible impurities in water that could make it more corrosive, and are removed by a strong magnetic field, but they can also be removed much more simply and cheaply by ordinary filtering. There are varieties of "health-giving" water that contain over 85% oxygen, or that are saturated with electrons, and so on and so forth. Of course, *all* water contains over 85% oxygen, and is saturated with electrons. Finally, I must once again (compare Chapter 4) mention homoeopathic remedies. These are formulated on the absurd principle that the more dilute a substance is, the higher its activity. So the most potent homoeopathic remedies are those that had been diluted to such an extent that they actually retain nothing but water. Defenders of homoeopathy argue that maybe the water retains a memory of the substances that have been dissolved in it. Well, actually, water does have a memory, as we know from very sophisticated measurements using the technique known as nuclear magnetic resonance. Unfortunately, this memory only lasts less than one billionth of a second. None of which has hitherto stopped the UK's financially hard-pressed National Health Service from funding homoeopathic hospitals, and at the time of writing their future remains the subject of intense political debate.

What Makes Life Possible

We have already mentioned that water is not the only substance that takes part in hydrogen bonding. We can classify molecules, or even particular regions within large molecules, as *hydrophobic* or *hydrophilic*. Hydrophilic (water loving) molecules attract water, either through electric charges or, more commonly, through hydrogen bonding. Hydrophobic (water fearing) molecules do not. Hydrophilic molecules can interact with each other, as well as with water. Langmuir introduced the idea of hydrophobic forces, which were supposed to draw hydrophobic molecules closer to each other. This does indeed happen, but in a rather indirect way. In a mixture of hydrophilic and hydrophobic molecules, the hydrophobic molecules, by their nature, are excluded from the network of hydrophilic interactions. As a result, they end up being pushed towards each other, rather like people at the wrong social gathering, who end up talking to each other, not because that is really what they want to do, but because they are excluded from the interacting networks forming around them.

Protein molecules are polymers built up from long chains of amino acids, and some amino acid groupings within the protein are hydrophobic, while others are hydrophilic. So protein molecules adopt very specific shapes, which bring the hydrophilic regions into contact with each other. The surface of a protein molecule contains a definite pattern of sites that can take part in hydrogen bonding to other molecules. Such patterns control the specificity of enzymes, which catalyse reactions in their target molecules but not in others, while direct pattern matching gives rise to the formation of antibodies, as in allergic reactions, and in the development of immunity to infections. The famous double helix of DNA is also held together by hydrogen bonding, which ensures that the information-carrying groups of each chain have the right partners on the other. So metabolism and reproduction are both possible only over the temperature range where groups of hydrogen bonds are stable enough to persist, but not so stable that they cannot be broken down and re-formed as necessary within living cells. This leads to the expectation that the range of temperatures over which life as we know it is possible, is more or less the same as the range of temperatures over which water is a liquid, an assumption that guides current thinking about the possibility of life elsewhere in the universe.

To Sum Up

Water owes its many strange properties to hydrogen bonding, the name given to the interaction between the polar O–H bonds in one water molecule, and the electron lone pairs on its neighbours. This imposes an open structure in ice, partly broken down in the liquid, which is why ice floats. It also accounts for the high boiling point and high surface tension of water, and why it is such a good solvent and wetting agent. Hydrogen bonding in proteins and other biologically important molecules controls molecular recognition, enzyme specificity, and nucleic acid replication. There are a few other substances where the liquid is denser than the solid. One of these is silica, silicon dioxide, which is present in granite and accounts for its tendency to solidify as it comes to the surface, instead of flowing like basalt. Thus the consequences of these special bonding situations range from the topography of landscapes to the geopolitics of trade routes to the chemical basis of life.

15 The Sun, The Earth, The Greenhouse

"To be clear, I believe in evolution and trust scientists on global warming. Call me crazy."

Jon Huntsman, former Governor of Utah,
former US Ambassador to China,
US Presidential hopeful at time of writing.

Something very disturbing is going on when a distinguished candidate for the most important job in the world has to apologise for agreeing with the overwhelming scientific evidence on these matters. And I would rather he had spoken of "accepting" the fact of evolution, rather than "believing" in it; evolution has been part of any reality-based worldview for almost a century, but that is another story. Hopefully, in the unlikely event that Jon Huntsman ever reads this chapter, he will emerge with his understanding of the reality of global warming based on something more than trust.

All Things Glow

The coldest emptiness of outer space glows in the microwave region, corresponding to a temperature of $2.725°C$ above absolute zero, the pale shadow of the light in which the universe was bathed when first it was cool enough to make atoms. The bar of an old-fashioned radiant electric fire glows red, and the surface of the Sun (I will explain what I mean by "surface" later) glows yellow-white hot. The surface of the Earth glows in

226

the infrared with an average temperature of around 15°C. But if you look at the emission spectrum of the Sun, you will find relatively dark lines on it, corresponding to the electronic absorption spectra of elements in the Sun's outer atmosphere, and if you look at the emission spectrum of the Earth, you will find relatively dark bands, corresponding to the infrared absorption spectra of the greenhouse gases.

The Sun

The Sun's core, which extends from the centre for about a quarter of the total radius, has a temperature of 13.6 million degrees Celsius and a density up to 150 times that of water. It is here that hydrogen is converted to helium, the energy released being radiated out in the form of gamma rays. The rest of the Sun, up to its visible surface, is a *plasma*, which is a state of matter consisting of cations (mainly protons, since hydrogen is the most abundant element present) and electrons, moving independently of each other. Time and again, over the space of a few millimetres, this plasma absorbs and re-emits the energy streaming outwards from the core. Since each layer is a little bit less hot than the one beneath it, the energy is passed on in   the form of electromagnetic radiation of slightly longer wavelength, until finally it reaches what we see as the surface. Unlike the surface of the Earth, this "surface" does not mark a sudden drop to lower density, but is the level at which the temperature is low enough for atoms to be able to hold on to their outer electrons. It is simply the layer where the plasma gives way to predominantly non-ionised gas. A plasma can absorb (and re-emit) light of any frequency, but an uncharged gas can only do this if the frequency corresponds to a change between definite energy levels.[1] So above the surface, the Sun's atmosphere becomes transparent at almost all wavelengths, and what we see is the surface itself, glowing yellow-white hot at 6000°C. Actually, the density of matter in this region of the Sun is so low (10^{13}–10^{14} particles/cm^3, compared with around 2.4×10^{25} particles/cm^3 in the Earth's lower atmosphere) that light emitted even from a depth of some hundreds of kilometres is able to escape, and instead of thinking of a single surface, we should think of a layer, the *photosphere*.

"We can never know anything of their chemical or mineralogical structure". So said the philosopher Auguste Comte in the 1830s, speaking of the stars. As we saw earlier,[2] he was wrong, and the investigations that would prove this were already well under way. Recall that (in 1672), Newton had shone a beam of sunlight through a glass prism, breaking it down into a *spectrum* with all the colours of the rainbow. Wollaston in England in 1802 and later Fraunhofer in Germany in 1814 repeated this work using the high-quality glass prisms that were becoming available, and noticed the existence of dark lines in that spectrum, still known as "Fraunhofer lines". Over the next decades, various scientists observed that metals and indeed elements in general when sparked or intensely heated gave out light at specific frequencies, something that could be used for analysis of materials, and that these frequencies matched the frequencies of the Fraunhofer lines.

By the 1850s, it was realised that this gave a way of doing exactly what Comte had called impossible. The presence of a particular set of lines was evidence for the corresponding element, and this was as true for the outer atmosphere of the Sun as it was for a mineral specimen here on Earth. Even more so if anything; the existence of the element helium was inferred from a dark line in the Sun's spectrum some 14 years before its first terrestrial detection, in lava from Mount Vesuvius.

We would have to wait until the 20th century for Planck and Einstein to discover the relationship between frequency and energy, and for Bohr to explain the lines in terms of discrete atomic energy levels, but this had no effect on their diagnostic significance.

In present-day terminology, the Sun's photosphere generates a continuous thermal emission spectrum. It glows because it's hot, and the spectrum is continuous because the plasma (and the hydride ions) can absorb and emit at any energy. Superimposed on this, we see the line absorption spectra of elements in the solar atmosphere, and these match the line emission spectra we can obtain in the laboratory.

I have referred to "line absorption spectra". It would be more accurate to speak of lines of less intense emission. Light of the frequency that matches a particular element is absorbed and re-emitted many times in its passage through the Sun's outer atmosphere, and its final intensity corresponds, crudely, to the temperature of the region from which the light is finally emitted to space. This is lower than the temperature of the photosphere, corresponding to less intense emission.[3] The distinction between absorption and less intense emission will turn out to be extremely important when we discuss the greenhouse effect on Earth.

One no doubt trivial effect of the existence of the dark lines is that the photosphere is slightly hotter than it would have been without them. The reasoning is very simple. Ignoring short-term fluctuations, and very long-term trends, the total amount of energy reaching the photosphere must be equal to the amount that leaves it. But the dark lines correspond to energy that has been prevented from leaving, so they must force the photosphere to adopt a slightly higher temperature, to compensate for less efficient final emission at the absorbed-reemitted frequencies.

The Earth

We turn now to the situation on Earth. Nearly all incoming light from the Sun passes straight through the Earth's atmosphere. An important exception is light in the far ultraviolet, which is absorbed by the ozone layer and, incidentally, heats the stratosphere in the process, but we can ignore this for our purposes. This light hits the surface of the Earth, where around 70% is absorbed, while 30% is directly reflected back into space (the Earth is therefore said to have an *albedo* of around 0.3).

So the Earth is warmed, and glows with the appropriate thermal radiation. Since the Earth's average temperature is around 15°C, or 288 degrees above absolute zero, as opposed to 6000°C for the Sun, the Earth glows in the infrared rather than in the visible. On average, the amount of energy reaching the Earth's surface must equal the amount of energy finally leaving it, and this balance, apart from a small contribution through geothermal heat from the energy of radioactive decay, is what determines the Earth's surface temperature. If there is not a balance, then in the short term the Earth either gets hotter (if more energy stays than leaves) or colder (if more energy leaves than stays), until a new steady state temperature is reached.

The Greenhouse

The atmosphere is transparent to visible light, but not to specific regions of the infrared, where some gases, the *greenhouse gases*, absorb, converting infrared light energy into molecular vibrational energy. Oxygen, nitrogen, and argon are transparent throughout the infrared, but carbon dioxide, methane, and water absorb (and, when energised, re-emit) specific frequencies of infrared light. Very importantly, while the atoms in the Sun's atmosphere absorb and re-emit sharp lines, the greenhouse gases absorb and re-emit over a range, giving band spectra, because of additional effects connected with molecular rotation.

To understand what happens next, we need to consider the structure of the Earth's atmosphere. The greater the altitude, the lower the pressure, because the pressure at any height is due to the weight of the air above it, which is why Himalayan mountaineers need extra oxygen, and why cake recipes from Massachusetts need to be modified in Montana. Nor will anyone who has taken a funicular chairlift need reminding that higher altitude leads to lower temperature. There are two apparently quite different ways of looking at this, but they come to the same thing. We can say that molecules simply lose kinetic energy, and therefore temperature, as they rise against gravity. Or we can think of air rising, expanding because of the lower pressure, and cooling because it is doing work on its surroundings (the opposite of the effect that makes the pump heat up when pumping up a tyre). So temperature decreases with height, through what is known as the troposphere, for a distance of some 4.5 miles at the poles or 10 miles at the equator. Above this height (the tropopause) we enter the stratosphere, where temperature begins to increase with height because of the absorption of ultraviolet by the ozone layer, along with further effects that are not important to this discussion.

Now consider what happens to the infrared radiation leaving the Earth's surface. At frequencies where the atmosphere is transparent, it will stream straight out into space. At frequencies where greenhouse gases absorb, that much energy will be captured, usually quite close to the surface, and is among the processes heating the lower atmosphere. But greenhouse gas molecules emit as well as absorb, so some of the absorbed energy will be deflected downwards, and some re-radiated upwards, where the process will repeat itself. Ultimately, some of this energy will percolate up to a level where it can escape into outer space. Since this escape level is at a lower temperature than the surface, the amount escaping is less than it would have been if the greenhouse gases had not been present, and the surface must be warmer to compensate. The result is that for a steady state to be established, the Earth's temperature must be some

33°C higher than would otherwise have been the case. To put it slightly differently, in addition to being warmed directly by sunlight, the surface is warmed by the infrared light re-emitted downwards by greenhouse gases, much as an actual greenhouse is warmed by infrared light reflected downwards by the window materials. None of this, by the way, is contentious.

Now consider further, what is bound to happen if there is an increase in the concentration of a greenhouse gas. The chance of one of its infrared photons being absorbed will be greater at any level, and the average level from which final escape occurs will be correspondingly higher. With the exception of some small regions of the spectrum where the escape layer is already at or very near to the tropopause, this means that the escaping emission will be coming from a layer where the temperature is that much lower, the total amount of energy escaping from the Earth will be reduced, and the surface of the Earth will adopt a higher temperature to restore the balance.

There has of course been a dramatic increase in the amount of carbon dioxide in the atmosphere over the past century and a half, related to the use of fossil fuels (and to a smaller extent to the use of limestone in cement manufacture), so that it now stands at about 40% higher than in preindustrial times. There has also been an increase in the amount of other greenhouse gases, such as methane and nitrous oxide, mainly as the result of more intensive agriculture, and the use of synthetic nitrogenous fertilisers. Most of this increase has taken place in the past 50 years, during which there has been a relentless upwards tendency in temperature. This is global warming. To be more specific, there is a direct causal relationship between increased greenhouse gases resulting from human activity, and the increase in temperature. This is *anthropogenic* global warming.

Three complicating factors deserve special mention. Firstly, water vapour. This is an amplifying factor, rather than an independent driver. If the Earth's temperature is pushed upward by any cause, the atmosphere will hold more water vapour, enhancing the greenhouse effect, and tending to produce a yet further increase. So we have a positive feedback loop. The matter is further complicated by cloud formation, which can enhance or reduce the greenhouse effect depending on exactly where the clouds form. Like greenhouse gases, they can absorb and re-emit infrared, enhancing the effect, but they also increase the Earth's albedo by reflecting light back into space before it arrives at the surface. The overall effect of clouds is the most uncertain aspect of greenhouse science. There is even a small but vocal group of climatologists who claim that clouds will provide negative feedback, mitigating the effects of anthropogenic global warming, but this is very much a minority point of view, and it seems more likely[4] that clouds will make things worse. Secondly, there is global dimming, as a result of industrial pollution. This was a large enough effect to override and indeed reverse the general warming trend for some 20 years in the middle of the last century. The passage of clean air acts removed the worst of this, whereupon, as already predicted at the time, temperatures resumed their upward movement. Finally, there is soot, not a gas but also a product of human activity, and of wildfires made more common by that activity. Soot is black and snow is white, so soot traps far more light energy than snow, causing serious melting in the Himalayas and Tibet.

Two other associated effects also deserve discussion. While melting sea ice has almost no effect on sea level (thanks to Archimedes' Principle), the melting of ice sheets over land could have a very serious effect indeed, and so, eventually, will the warming and expansion of the water in the oceans. And an increase in atmospheric carbon dioxide leads directly to an increase in the acidity of the oceans, and hence in the solubility of the calcium carbonate from which coral, plankton, and other forms of marine life build their shells.[5] This is one of the least discussed, but most troubling, aspects of the entire process. Quite apart from their role at the bottom of the oceanic food chains, photosynthetic plankton remove vast quantities of carbon dioxide from the atmosphere, and disruption of this process then introduces a positive feedback into both acidification and climate change.

Fire, Flood, and Famine?

2010 was the hottest year and came at the end of what was certainly the hottest decade since record-keeping began, and probably the hottest for many thousands of years. Over the past few years, drought in Russia and elsewhere has sent grain prices soaring. We have seen cities from Moscow to Sydney experiencing record heat, and being shrouded in smoke from burning forests and smouldering dried out peat bogs. Monsoons fed by unusually high evaporation from the warming oceans have brought devastating floods in Pakistan and China, while as I write the 2011 hurricane season in the US shows every sign of unusual severity, Russia is plagued with forest fires for the second year running, and the Horn of Africa is tortured by drought and famine. Aren't these apparently ever more frequent natural disasters the result of global warming?

Regarding any one of these events, there is only one honest answer. Maybe. But as time goes by, and as we consider their totality, that "maybe" is turning into a "probably". We were warned that such events would happen with increased frequency, and indeed they have. But attributing any specific event to so general a cause is going beyond the evidence. Like loaded dice, global warming changes the odds, but individual outcomes remain unpredictable.

Regarding overall warming, drought, changing rainfall patterns, declining crops, and the additional effects of acidifying and slowly rising oceans, the melancholy answer is that yes, these things are happening now, driven by our greenhouse emissions, as long since predicted. After all, the Intergovernmental Panel on Climate Change has been reporting since 1990, the first three reports having been edited by Sir John Houghton, former Professor of Atmospheric Physics at Oxford, whom we met in Chapter 7. (Sir John's book *Global Warming, the Complete Briefing*[6] is an excellent overall review of the subject.)

But What About...?

Apart from the special case of biblically inspired objections to the opening chapter, this is the only part of the book where I can expect large numbers of readers to think that what I'm saying is not true. This is no accident. There is a multi-million dollar publicity machine dedicated to creating the impression, in the minds of the public, that the science is not completely settled and that, therefore, no political action is necessary. Not surprisingly, the machine is funded by those with most to lose, both financially and in terms of ideology, and its conclusions are embraced by those most opposed to political action of any kind. The techniques are similar to those used for so many years by the tobacco companies to sow doubt concerning the relationship between smoking and lung cancer, and indeed some of the same people are involved, including Fred Singer, who much earlier in his career (see Chapter 7) had done useful service in improving techniques for measuring the health of the ozone layer. (For a full discussion of this publicity machine and its tactics, see *Merchants of Doubt*, by Oreskes and Conway.)[7]

To say nothing of the way the question has been politicised, especially in the United States, as if the truth about Earth's climate were a matter of ideology.

After all, weren't we told in June 2011, by James Delingpole, whose work is featured by the (London) *Telegraph* and *Times*, that "a new Ice Age is on its way. In what has been described as 'the science story of the century', heavyweight US solar physicists have announced that the sun is heading for a prolonged period of low activity"?[8] No matter that the source doing the describing was Fox News, or that what is being described is low sunspot activity, or that this much had already been surmised a year ago, and the effect estimated at −0.3°C, a fraction of the anthropogenic effect.[9] If you pay attention to the *Times* or *Telegraph*, let alone Fox News or the *Daily Mail*, and if you make the reasonable but sadly unwarranted assumption that they only use sources who know what they are talking about, you will be led to conclude that I, poor misguided soul, am sadly out of date.

Such is the volume of disinformation, that it may be helpful if I run through some of the more commonly used arguments. Like creationists' arguments (and indeed we are often talking about the same individuals), they have been refuted many times, but this has no effect at all on the rhetorical intensity with which they continue to be put forward. Many readers will have seen all of this before, and can simply skip to the next section. There is an excellent discussion of all these arguments, with links to the original literature, at http://skepticalscience.com/.

1. *Carbon dioxide is only present at around 380 ppm (parts per million, as a fraction of the number of molecules) in the atmosphere, so it can't have any major effect.* Nonsense. This argument reveals such a total failure to understand what is going on, that it is difficult to believe in the good faith (or alternatively the good sense) of those who promote it. We are talking about several miles thickness, so it mounts up. Besides, low concentration does not mean low effectiveness, especially when we are talking about the absorption and emission of light. One might as well say that the ozone layer must be unimportant, because the total amount of ozone is so small.

(Come to think of it, the same people who campaigned against the 1987 Montreal Protocol to protect the ozone layer are now vocal in their rejection of the science of climate change.)

2. (Almost the opposite of the first argument) *There is so much carbon dioxide in the atmosphere already that it traps all the infrared it possibly can, so that its effect is saturated.* This argument sometimes comes accompanied by copious equations, which help to conceal its fallacies. As I explain above, what matters is not the level at which the infrared is initially trapped, but the level at which it is finally released. And although at the peaks this is already so high in the stratosphere that indeed added carbon dioxide will make little difference, the same does not apply to the broad "wings" on either side of the absorption bands.

3. *Water is a more important greenhouse gas than carbon dioxide.* In a sense, this is true, but not in the sense that climate change deniers want you to think. Water vapour does not act independently of carbon dioxide. On the contrary, it amplifies its effect. This is because the concentration of water vapour in the atmosphere is fixed by surface conditions, so that unlike carbon dioxide, it is not an independent driver of change, but magnifies effects that are already there. The more warming, the more water vapour, amplifying that warming. Another positive feedback loop.

4. *It's the Sun.* No it isn't. The eleven year solar cycle is detectable as a minor wiggle on the general upward trend. Night-time warming has been greater than daytime, and the upper atmosphere has cooled while the lower atmosphere has warmed. This is as expected from an enhanced greenhouse effect, while variations in solar output would have produced the opposite. Moreover, the last decade of rising temperatures has taken place despite a weak sunspot cycle, which is thought to have a cooling effect.

5. *The Earth's climate has always fluctuated, and always will, so it is foolish to try to do anything about it.* That's just like saying that there have always been outbreaks of disease, and always will, so it is foolish to practice hygiene or vaccination. Looking at the argument a little bit more closely, we find that indeed there have been far larger fluctuations, but they have generally been far slower, allowing time for species to adapt. Nor have species always been able to adapt very successfully; after all, there are five major extinctions in the geological record, and it is thought[10] that rapid changes in the atmosphere may have been responsible for at least some of these.

5. *Human emissions are puny compared with those from volcanoes.* FALSE. This so-called argument keeps cropping up, and was, deplorably, featured in a programme, *The Great Global Warming Swindle*, that was shown on Channel 4 Television in the UK. The reality is[11] that carbon dioxide from human activity exceeds carbon dioxide from volcanoes by a factor of around 100 to 1.

6. *The Earth has been warming for several thousand years, as it emerged from the last major ice age.* Sometimes, the same argument is used with reference to the "little ice age" of the 16th to 19th century. This boils down to saying that the Earth is now warmer than it once was, because it used to be cooler. True, but not very informative. Moreover, the current warming trend is much faster than what came before, and cannot be simply dismissed as part of the same trend.

7. *What about the Medieval Warm Period, when they used to grow grapes in Iceland?* Oh no they didn't, nothing like. They did grow them in southern England, but now you can, just about, grow grapes in Scotland.

8. *What about the urban heat island effect?* This was a very popular argument for a while. The suggestion was, that centres of population generate excess heat (this is true), and that therefore the observed increase in temperatures was just the result of local activity, rather than global warming. However, the upward trend was completely unaffected (in fact, surprisingly, appeared to be a little bit greater) when the records from sites near cities were removed from the calculation.

9. *But what about Climategate?* Here's what. Thousands of e-mails from scientists at the Climate Research Unit at the University of East Anglia, and their colleagues elsewhere, were illegally downloaded, and out-of-context snippets passed on to the press, two weeks before the 2009 Copenhagen climate conference, in an attempt to suggest manipulation and misrepresentation of data. Further stolen e-mails were released on 22 November 2011, days before the Durban climate conference. Seven separate major investigations on two continents (UK House of Commons Science and Technology Committee, UK Independent Climate Change Review (the Russell Report), UK International Science Assessment Panel (the Roxburgh Report), and inquiries in the US by Pennsylvania State University, the Environmental Protection Agency, the Department of Commerce, and most recently, reporting in August 2011,[12] the National Science Foundation) found nothing worse than minor occasional sloppiness, but this has not

killed the theory that climate change science is one vast conspiracy. Such a conspiracy would have to involve thousands of scientists who are experts in the area of climate studies and large scale computer modelling, over dozens of separate laboratories, but conspiracy theories, as many examples show, are invulnerable to reason, and the political posturings that thrive on repetition of such allegations are invulnerable to evidence.

A more sophisticated version of the argument says that we shouldn't trust climatologists to advise us about climate, because in the nature of things they have a vested interest in the importance of the subject. This is a little bit like saying that we shouldn't trust metallurgists to advise us about metals, because in the nature of things they have a vested interest in the importance of these materials. Of course climatologists think that climate is important; that's why they have chosen to spend their professional lives studying it.

10. *Some 19th-century measurements show carbon dioxide levels even higher than today's.* True, but these are clearly in error. Carbon dioxide levels are now measured using infrared absorption, but in the 19th century they were assessed from the acidity of the atmosphere. The best such "direct chemical" assessments agreed well with those found in ice core layers laid down at that time, and the high outliers can readily be explained by local atmospheric pollution, especially pollution by acidic sulphur dioxide from coal burning.

11. *The increase in carbon dioxide is natural.* Not true. We can tell this from the isotopic composition of carbon dioxide in the atmosphere and how this has varied over time, as captured in the wood of tree rings. This shows two things. The amount of ^{14}C generated by cosmic ray activity has been progressively diluted by the larger total volume in circulation,[13] and there has been a small but measurable decrease in the atmospheric abundance of the minor isotope, ^{13}C. This ^{13}C depletion tells us that the increased volume comes from material that has been biologically processed. We can also estimate the amount of carbon dioxide emitted by burning fossil fuels. It turns out to be about twice the amount that has been added to the atmosphere, the rest having been absorbed by the oceans, moving them towards acidity.

12. *1998, or 1937, or 2005, or whatever, was the hottest year on record.* This may arguably be true of one particular location, but is certainly not true globally. It is just an example of cherry-picking; there are ups and downs

in these things, as well as local variations. So the purely local record highs of the dust bowl years in the central United States stood for several decades, and globally 1998 was not surpassed until 2005, and 2005 not surpassed until 2010. So between 1998 and 2005, the enemies of climate science claimed that the planet had been cooling since 1998, between 2005 and 2010 they claimed it had been cooling since 2005, and if, as is quite likely, 2011 comes in slightly below 2010, they will no doubt tell us that the cooling trend has been resumed.

13. *It isn't really warming. The most recent decade has been on average no hotter than 1998.* Actually it has been, but not by much. Like the last objection, this is cherry-picking, in this case by comparing a ten year average with a year selected because it was, in its decade, unusually hot. It is true that there has been less warming since 1998 than predicted, but that is for a very specific reason; the vast increase in coal burning and sulphur dioxide emissions from China's rapid and under-regulated economic growth, causing acid haze and global dimming.[14] This haze will wash out over the next few years; the carbon dioxide emitted will stay up there.

14. *Glaciers and ice sheets are recovering.* No they're not. The Heartland Institute, which exists to promote the ideology of free enterprise and is shameless in the tendentious arguments that it uses to this end, recently made great play of the fact that in one month in 2004, the extent (just the area, not the volume) of the Arctic ice sheet was marginally greater than it had been 10 years earlier. This is cherry-picking with a vengeance, and to put it in context the area is now some 10% less than it was in 1994, and 20% down on the levels that held steady throughout the first half of the 20th century. Sea ice volume, a better indicator than area, hit a minimum in 2007, probably surpassed in 2010, and has been declining at a steady rate of some 2800 cubic kilometres per decade since 1979.[15] Similarly, it is possible to cherry-pick one's glaciers. Glaciers in some valleys in the Karakoram range are growing, but this is actually a result of warming, leading locally to more precipitation, and is a minor effect compared with the overall loss.

15. *We really ought to be worrying about the next ice age.* This is an old claim that is suddenly back in fashion, and some of the climate scientists now concerned about global warming were indeed, back in the 1970s, warning us about global cooling. There are two different arguments being muddled up here. Indeed, another ice age is expected, some time between 5000 and 10,000 years in the future or perhaps longer; this has nothing to do

with present policies. And there were voices (even then, a minority) warning of global cooling in the 1970s, as the result of pollution and global dimming. Since then we have had the clean air acts and, predictably, the anthropogenic cooling has subsided. If there is any moral at all to this story, it is that climate science works.

16. *The implications of climate change are (for some people) politically unacceptable.* They have led the US National Academy of Sciences[16] to commend "strong federal policies" and "strong U.S. engagement in international-level response effort," suggestions that to some are anathema, for reasons that have nothing to do with science.

Tough. Reality does not depend on what we do or do not happen to find acceptable. If we reject an argument because it has consequences that are untrue, that is logical thinking. If we reject it because the consequences are unpalatable, that is *wishful* thinking, and those who make their decisions on the basis of wishful thinking are inviting retribution from reality.

For what it's worth, I would be delighted if there were any valid arguments against the current scientific consensus on climate change. I have been looking for such arguments for over 20 years, without success. I have grandchildren, and am concerned about the kind of world that they will be growing up in, and climate change and its implications are the gravest of these concerns.

Uncertainty and its Implications

The last ditch argument against action on climate is in some ways the most difficult to counter. There is genuine uncertainty about the size of the effect, so does it really make sense to divert real present resources against an ill-defined future threat? Such reasoning is seductive but specious. There is indeed uncertainty, regarding just how serious global warming will be under various scenarios, but this uncertainty is a reason for more circumspection, not less. A prediction of $3°C \pm 2°C$ is far worse than a prediction of $3°$ with no uncertainty. $3°$ will be uncomfortable, but $5°$ would threaten our civilisation.

It is because of uncertainties that we insure our houses, and maintain our armies. It is because of their interest in evaluating uncertainties that insurance companies have been among those most attentive to climate change research, which they started funding in the 1970s. If things may be rather less bad than we predict, they may also by the same token be a great deal worse. In the words of Margaret Thatcher, "[T]he need for more research should not be an excuse for delaying much needed action now. There is already a clear case for precautionary action at an international level." This is one of the few occasions when I found myself in complete agreement with the Iron Lady. What she said was right then, and it is right now. She was speaking, by the way, in 1990.[17]

To Sum Up

Sunlight is the way it is, because of the way that the energy generated in the core is reprocessed before being finally emitted from the photosphere close to the surface. Earth's energy balance depends on the Earth emitting energy in the form of infrared light to balance the heat we get from the Sun. Greenhouse gases cause some of this energy to be emitted from higher levels of the atmosphere, which are cooler, so that the emission is reduced. So energy balance requires the surface to be at a higher temperature than it would have been, if these gases had been absent. None of this is controversial.

It has become increasingly clear, beyond all reasonable doubt, that the increase in anthropogenic emissions, especially of carbon dioxide, is enhancing the greenhouse effect and driving a steady increase in temperature, which can be expected to have extremely serious consequences. Many arguments have been devised by those who wish to avoid this conclusion, but none of them will stand examination.

16 In The Beginning

We ourselves are part of the natural world that we study. Every carbon atom in our body has been captured by photosynthesis from the atmosphere, every carbon atom on our planet was fashioned in some long-exploded star.

In the Beginning ...

But was there a beginning? In the almost 50 years since the "Big Bang" theory gained general acceptance among scientists, there has been much speculation about what such a beginning could possibly mean. Could there be such a thing as a time "before" the Big Bang, or did time and space only come into being when there was matter to occupy them? Was "our" Big Bang unique, or is ours one of a perhaps unimaginably vast number of independently existing universes? Are new universes generated inside black holes? If so, do these daughter universes obey exactly the same physical laws as their parents, or could there be small variations, giving rise, according to one theory,[1] to a kind of Darwinian evolution, favouring universes that can give rise to offspring? Can the earliest state of the universe be understood in terms of string theory, or in terms of its close relative, brane (membrane) theory, according to which it could have arisen from a collision between branes? Or is our own Universe, which on current theories seems destined to continue expanding forever, the last of an indefinitely long series, each one generated by its own Big Bang, only to collapse at the end in a Big Crunch? Is there any way, even in principle, that we can find an answer to these questions? Observations outside our own Universe are impossible, by definition, but could some of these multi-universe theories be more more compatable than others with the existence of a Universe like our own? And, lacking as we still do a theory that can embrace both quantum mechanics and General Relativity,[2] how can we begin to describe the state of our Universe immediately after its birth, when it was so tightly compressed that gravity was as strong as the forces that now hold the atomic nucleus together?

There are people of great ability addressing these issues, but I must content myself with a much more modest question.

What has happened in that part of the Universe that we can observe, since one trillionth of a trillionth of a second after it came into being?

An Inconstant Universe

A century ago, it was believed that the size of the Universe was unchanging. Why imagine anything else? Indeed, so strong was the appeal of the idea of an unchanging Universe, that when the equations of General Relativity predicted expansion, Einstein added a "cosmological constant", or fudge factor, to hold it steady. Then, in the late 1920s, the American astronomer Edwin Hubble[3] suggested testing the expanding Universe theory using the light from distant galaxies. He was able to do this since he could recognise the characteristic emission lines of the atoms contained in them, from their relative spacing, and found that the actual frequencies were *red shifted*. The greater the distance, the greater the shift, according to a linear relationship now known as Hubble's Law, although it had already been independently formulated by the Belgian theoretician (and priest) Georges Lemaître.[4] Subsequent decades have confirmed this observation, to greater and greater distances. From the fact that the spacing patterns are recognisable, we can confirm an extremely important assumption, one that most of us make without thinking; the laws of nature must be the same throughout the region of space that we can sample. They must also be constant over the full extent of time that we can sample, because light takes time to reach us, so that as we look out into space we are also looking backwards in time. We do not see the Sun as it is now; we see it as it was eight minutes ago, because its light takes eight minutes to reach us. The most distant observed galaxies are over 12 billion light years away, meaning that we are seeing them as they were more than 12 billion years ago. There is some evidence that there may be tiny variations in the pattern, corresponding to a difference in the laws of physics[5] in the sixth decimal place. This is at the limits of the accuracy of observation, but the very possibility has aroused enormous interest.

The red shift means that distant galaxies seem to be rushing away from us, and the speed at which they do this increases in direct proportion to their distance. Or, as physicists prefer to put it, their distance from us is increasing because new space is being created in between them and us. This makes us sound privileged, at the centre of things. However, a little bit of mathematics shows that every observer will get this same impression.[6] It also means that if you run the movie backwards, everything starts out from the same place at the same time. In 1929 this led Lemaître to propose[7] an early version of the Big Bang theory, according to which our entire Universe began in a single hot atom, and using the more accurate measurements that we now have of how red shift varies with distance, we can calculate that this happened some 13.7 billion years ago.

For many years this Expanding Universe theory, as it was then called, was one of two under serious consideration. The other one was known as Continuous Creation, or Steady State. This was based on the very attractive idea that any one large enough chunk of space-time would look very much like any other. But the galaxies are flying apart, decreasing the density of the Universe. In Steady State theory, this was balanced out by new hydrogen atoms simply popping into existence at a rate just sufficient to balance this effect, which would have been far smaller than anything we could hope to observe directly. The chief exponent of this theory was the British astronomer Fred Hoyle, who was also responsible for helping to develop the theory that stands to this day of how these hydrogen atoms would form stars, and how reactions within stars, as we shall see, gave rise to all the different elements that exist today. It was Hoyle who coined the expression "Big Bang", semi-facetiously, in a radio interview back in 1950, and the name stuck.

The tragedy of science, said Darwin's friend and ally Thomas Huxley, is this: an elegant hypothesis slain by an ugly fact. I can remember my sadness when this tragedy overtook the Steady State theory. Remember that the fundamental principle of this theory is that one chunk of space-time looks much like any other. This would mean that, as we gaze outwards in space and backwards in time, what we see should not change on average with distance. But it does. Further away, the galaxies are closer together, as they would have been, going back in time, on the Big Bang theory, and there are other differences indicating that we are looking at a younger Universe. This is simply not consistent with the Steady State theory, and so, despite its elegance, it must be rejected.

Besides, in 1965, a dramatic discovery was made, that turned out to match a prediction made much earlier about the consequences of a Big Bang. Two engineers working at Bell Laboratories were trying to solve the problem of cutting out noise in radio communications. But try as they would, there was always a certain microwave frequency hiss in the background, persisting with equal intensity in all directions. Nothing at all could get rid of it, although they tried everything they could think of, even including removing pigeon droppings from their antenna. When they examined how the intensity of this hiss varied with frequency, they found that it corresponded to the thermal radiation from an object at a temperature of 2.7°K, 2.7° above absolute zero. The existence of this cosmic microwave background had in fact been predicted long before by George Gamow, a Russian physicist and friend of Niels Bohr who had managed to make his way out of the Soviet Union in 1933. It was not possible to really observe it properly, because of atmospheric absorption and other effects, until in 1989 NASA launched COBE, the orbiting Cosmic Background Explorer, but the Big Bang theory had won general acceptance long before then.

The Big Bang, and its Aftermath

Going back in time means concentrating together all the matter, and all the energy, in our Universe. This takes us to conditions of temperature and pressure far beyond our everyday experience, and even far beyond the most energetic objects that astronomers can observe, such as supernovae, pulsars, and the active galactic centres in which matter is spiralling inwards towards enormous black holes. The closest we can come towards observing these conditions, is by way of the most energetic atom–to–atom collisions accomplished within our most powerful particle accelerators, such as the Large Hadron Collider underneath the French–Swiss border. So we have a convergence between the study of the very smallest objects, the fundamental constituents of matter, and the largest object of which we can have any knowledge, namely the Universe itself.

One trillionth of a trillionth of the second after its beginning, the temperature of the Universe was so high that matter was broken down into its ultimate components, particles known as quarks. As the Universe expanded and cooled, and the temperature fell to a mere two trillion degrees or so,[8] these combined together to give protons and neutrons, squeezed as closely together as they are in an atomic nucleus. Further expansion and cooling, to one billion degrees after 100 seconds or so, led to the formation of separate protons and neutrons, as well as electrons, with protons and neutrons now fusing together to give helium nuclei,[9] and small amounts of deuterium and lithium, during the first minute or two. At this stage, energies were still far too high for these nuclei to be able to hold onto their electrons and give neutral atoms. All the matter in the Universe was in the state known as *plasma*, which interacts strongly with electric fields, including the oscillating electric field of light, making it completely opaque. We have to wait another 380,000 years, by which time the Universe had attained just under one thousandth of its present size, and had cooled to around 3000° above absolute zero, before temperatures were low enough for atoms to recombine with their electrons, and light could at last travel freely through space. This was the era when the background radiation was born, filling the whole of space, and progressively cooling with it, until it now corresponds to a mere 2.7° above absolute zero.[10]

A Star is Born, and Lives, and Dies

At this stage we had space almost uniformly filled with the same mixture of elements that had been generated in the minutes after the Big Bang. Such a situation is unstable to chance fluctuation. A patch of slightly higher density will have stronger gravitational attraction, will attract more matter towards it, contract, and become denser still. The energy released in this process will heat up the gas, increasing its pressure, which would bring the process to a halt, were it not that hydrogen, by now cool enough to form molecules, turns heat energy into vibrational energy, which is then emitted to cooler regions. This is the beginning of the process that leads in turn to the formation of galaxy clusters, of individual galaxies, and eventually, within the galaxies, of individual stars. Within the newly forming star, the gravitational energy released is converted to heat, but as Lord Kelvin had argued, the energy available in this manner would only be enough to account for a few tens of millions of years of energy output in a star like our own Sun. Eventually,[11] however, temperatures and pressures become high enough to overcome the electrostatic repulsion between hydrogen nuclei, and the process of "hydrogen burning", which produces energy by converting hydrogen into helium, begins.

The easiest way to keep track of what happens in these nuclear processes, is to focus on what is called the *mass number* of the different kinds of atom involved. This is simply the sum of the number of protons and neutrons in the nucleus. These two fundamental particles are reasonably close to each other in mass, and also quite close to the mass of the hydrogen atom, which is why the atomic weights of so many elements, which consist almost entirely of a single isotope, are very close to whole number multiples of that of hydrogen (hence Prout's hypothesis, which played such an important role in the discovery of the noble gases). Total mass number (what is sometimes called *baryon number*) is conserved in all processes. So is electrical charge, and this is achieved by having an atomic nucleus emit an electron (beta decay), absorb an electron from its surrounds (electron capture), or emit a positron, which is the positively charged antiparticle to the electron (beta plus or positron decay). Since light particle (*lepton*) number is also conserved, emission of an electron (lepton number 1) is accompanied by emission of an antineutrino (lepton number -1). The mass of an atom is very close, in "atomic mass units", to the mass number, but there is a slight difference,[12] reflecting the energy of formation of the nucleus. Small differences in mass correspond to enormous differences in nuclear binding energy, and these are the differences that drive the energy production of stars.

Fusion of two ordinary hydrogen nuclei (these of course are just single protons) leads to heavy hydrogen, or deuterium, with mass number 2, and deuterium then gives rise (by a rather roundabout route) to helium-4. In stars more massive than our own Sun, there is also a catalytic cycle, whereby carbon-12 gets converted to nitrogen-15, which reacts with a proton to give helium-4 and regenerate carbon-12. Where did that carbon-12 come from? See below.

By either pathway, conversion of hydrogen to helium occurs in a relatively stately manner across almost the entire lifetime of a star, during which it is said to lie on the *main sequence*, on a graph of brightness against temperature. During this period, the material making up the star is in equilibrium between two opposing forces, the force of gravity pulling it inwards, and pressure pushing it outwards, and part of this pressure is due to outwards streaming radiation. As the reaction proceeds, the hydrogen at the very centre of the star is progressively converted to helium, and the star becomes progressively hotter as the reaction zone moves outwards, until the supply of hydrogen is close to being exhausted. Towards the end, the process is taking place close to the surface of the star, where gravity is much weaker than at the core. The result is that the surface layers expand and cool, converting the star to a red giant. This entire process is expected to take some 10 billion years in the case of our own Sun, so that we are now, very roughly speaking, half way through it. Larger stars burn brighter and fiercer, and move through the main sequence in a shorter time, a fact that turns out to have considerable importance. A star three times as massive as our Sun will get through the main sequence stage in less than a billion years, while one of half the mass will take nearly 50 billion; remember that our Universe is still only some 13 billion years old.

What happens next depends on the size of the star. It is predicted that a star half the mass of our Sun will just collapse in under its own gravity, becoming a small "white dwarf", radiate away its energy, and fade to a "black dwarf". However, there are no such black dwarfs as yet, because our Universe is nowhere near old enough. Our own Sun will heat up during its collapse to around 100,000,000°, at which point processes called the *"triple alpha process"* and *"alpha process"* begin to happen. Of course, we cannot observe these directly within the stars, but we can infer what temperatures and pressures exist within stars from the requirement that pressure has to balance gravitational attraction, we can obtain vital information (such as the relationship between brightness and temperature; the greater the overall brightness, the higher the temperature) from traditional astronomy, and we can now study reactions between different kinds of nuclei in the laboratory using particle accelerators. This is the kind of information used to build up the following account, which has survived remarkably well since first put forward by Hoyle and his colleagues over 50 years ago.[13]

In the triple alpha process, three helium-4 nuclei (alpha particles) fuse together to form one carbon-12 nucleus. Even under the extreme conditions of pressure in the reacting star, there is no way in which three separate particles, all positively charged, are going to come together, so the reaction takes place in two stages. In the first of these, two alpha particles fuse together to make one beryllium-8 nucleus. This however is actually unstable, and flies apart again almost immediately. For the reaction to proceed further, it is necessary for the beryllium-8 to react within its brief lifetime with another alpha particle, to give carbon. For this to occur, Fred Hoyle reasoned that there must exist an excited state, or "resonance", of the carbon nucleus, within which the three reacting alpha particles would be held together for long enough to emit the surplus energy as a gamma ray. Some of this carbon-12 reacts in turn with another alpha particle, by the alpha process, to give oxygen-16 and smaller amounts of neon-20 and other, heavier elements. Temperatures will rise high enough to generate neutrons, which will be captured by the various nuclei present, as will hydrogen nuclei mixed back by convection currents into the reacting core, giving other elements and isotopes, including nitrogen, and the carbon-13 and the oxygen-17 and -18 that we met earlier, in our discussion of how a stalagmite functions as a recording thermometer and rain gauge. Towards the end of this process, which will take a mere billion years (there is less energy to be released by these reactions, than was released by the conversion of hydrogen to helium), the Sun will swell up once more, shedding some of the material formed in these later phases into interstellar space, before settling down to become a white dwarf.

The Death of Stars, the Birth of Elements

Such is the fate of the great majority of stars, all those with an initial mass less than about eight times that of the Sun. Larger stars will go through the same stages, but on a timescale of millions rather than billions of years. Continued gravitational collapse will then raise the temperature to billions of degrees, making further reactions possible (the bigger the nuclei, the larger the repulsion between their electrical charges, and the greater the amount of thermal energy needed to produce reaction). At the same time, when large nuclei fuse, smaller fragments can get detached and react further. Carbon burning gives rise to neon, sodium, and magnesium (mass range up to the mid 20s); neon burning to magnesium and silicon (mid-20s to around 30); oxygen burning to silicon and sulphur (low 30s); and silicon burning to the elements from titanium to zinc (mass numbers up to about 60). The temperatures required for silicon burning are so high, that they enable other reactions to take place, leading towards a thermodynamic equilibrium. Along this sequence, stability per mass unit is increasing, with the greatest stability around 60 (actually, at iron-56, which accounts for the relatively high abundance of iron on Earth). The elements being synthesised at this stage now contain more neutrons than protons, and as a result[14] the star generates energy in the form of neutrinos, extremely small particles that interact only very weakly with matter, so that unlike photons they can stream outwards unimpeded. Within the star, pressure is now dominated by a quantum mechanical effect, known as *electron degeneracy pressure*, which is the resistance of electrons to being squeezed into smaller volumes. Very simply put, the electrons resist being crammed into a smaller space, because that implies less uncertainty about their position, and therefore greater uncertainty about their momentum, which in turn requires higher kinetic energy.[15] Degeneracy pressure depends on density, but not on temperature, so that the enormous heat generated in the core does not lead it to expand. As result, reactions can proceed ever more rapidly. The star develops an onion-like structure, with silicon burning and equilibrium in the inner-most core, surrounded as you move outwards by shells rich in silicon and sulphur, then in oxygen, neon, and magnesium, then carbon and helium, then in helium and unreacted hydrogen from the outermost layers, with reactions taking place on the boundary between these shells. Meantime, increasing radiation pressure leads to an ongoing increase in the stellar wind, a phenomenon present in all stars and caused by the pressure of radiation on their outermost layers.

Matter in the iron core has reached its most stable nuclear state. It is no longer generating neutrinos or outward-streaming photons. If the mass of the iron core reaches a critical value, called the Chandrasekhar limit after the astronomer who long ago predicted its existence, gravita-tional forces compress the matter so strongly that the atomic nuclei are forced to swallow the electrons surrounding them, and the core becomes as densely packed as an atomic nucleus. This cataclysmic event releases huge amounts of gravitational energy, heating the overlying layers to the point where matter breaks up once more into its simplest components, and neutrons and neutrinos come streaming outwards. During this brief phase, the star is emitting light as brightly as an entire galaxy, while up to 10% of its total mass is converted into energy carried away by the escap-ing neutrinos. We have what is called a Type II or gravitational collapse supernova,[16] ejecting enormous amounts of dust into the surrounding interstellar space, together with a shock wave that spreads through the gas cloud, compressing it in places until it is dense enough for the process of star formation to start up once again. Another kind of supernova, Type I, is ignited when a white dwarf, consisting mainly of carbon and oxygen, is pushed to the brink of the Chandrasekhar limit by the accretion of mass captured from a companion. This gives rise to runaway nuclear reaction, and the eventual total disruption of the star, scattering its material into space. The supernova reported in August 2011 is of this kind. Such supernovae are of particular importance in observational astronomy. Simplifying somewhat,[17] since they all arise at the same mass limit, they should all emit the same amount of energy, and can serve as "standard candles" whose apparent brightness makes it possible to determine the distances of other galaxies.

But what about the elements heavier than zinc? These are formed while the supernova is at its most active, sending neutrons streaming outwards from its centre. These neutrons are captured by the atomic nuclei, causing an increase in mass number by one unit for each neutron captured. But if the neutron to proton ratio is too large within a nucleus, it will be unstable, and emit an electron (and an anti-neutrino) to move to a stabler arrangement. In a Type II supernova explosion, so many neutrons are generated that this decay process cannot keep up, and we have what is called the *r* process (*r* for rapid), pushing up the mass number as one neutron is captured after another, followed by eventual beta decay to the most neutron-rich stable nucleus that can be formed with that mass number. (Recall that in beta decay, a neutron within the nucleus is converted to a proton, with loss of an electron and — to balance light particle number — an anti-neutrino. So the mass number of the nucleus, the total number of protons and neutrons combined, remains the same). In less spectacular star deaths, including in due course that of our own Sun, neutrons arrive more slowly, allowing time for each nucleus to decay to a stable state before acquiring another neutron, and this *s* process gives rise to a different, less neutron-rich, set of final products. The *r* process would have been in operation from the time of formation and rapid collapse of the very first massive stars, while the *s* process would have made its appearance much later. Both *r* and *s* processes must be invoked in order to explain the relative abundances of the isotopes found in the Solar System, and inherited, as we shall see, from an earlier generation of stars.

It is calculated that there are around five supernovae a year in our own galaxy, but it is difficult for us to observe these because of the amount of dust in the galactic plane. We are, however, fortunate in having been able to observe Supernova SN 1987A, in the Large Magellanic Cloud some 170,000 light years away, the light from which arrived on Earth in 1987. The Large Magellanic Cloud is a small galaxy near to our own, and moving in its gravitational field. SN 1987A was a gravitational collapse supernova. The remains are surrounded by a triple ring of material, whose position and speed show that it was formed some 20,000 years before the explosion itself, and this is evidence that the star was pushed over the Chandrasekhar limit by the merging of two large stars in a binary system.[18] In the case of this supernova, we were able to detect the neutrino burst from the collapse, and to directly confirm the formation of the r process product, nickel-56, during the explosion, from the half-life of the specific emission associated with the decay of its daughter, cobalt-56, to stable iron-56.

The s process takes place less dramatically, over millions of years, in red giant stars. We can nonetheless be sure that it is taking place, from the chemical composition of these stars, as revealed by their spectra. For example, some of them[19] contain detectable amounts of the unstable element technetium, which has no isotope that could last for more than a few million years. Since the star itself is much older, the technetium must be forming inside it.

Now we can answer questions that at one time seemed unanswerable. Epicurus was wrong. Atoms have not existed forever. The lightest elements were formed in the minutes after the Big Bang, and all others have been fashioned and refashioned within stars. The Universe, or at least the observable Universe, is not unchanging. It is getting larger, as light from more and more distant galaxies finally makes its way to us. It is also getting more dispersed, as more space is created between the galaxies, and dustier, as more stars go through their lifes' trajectories and deposit some of the heavy elements they have made into interstellar space. Kelvin was right in thinking no force known to the science of his day could fuel the energy of a star over what we now recognise, despite Kelvin's protests, as geological time. Rutherford was wrong in thinking that radioactive decay would supply enough energy to make up the shortfall. Niels Bohr, never afraid to challenge established principles, suggested at one stage that the law of conservation of energy was somehow violated within stars.

He was wrong, too. The energy comes from the same processes of fusion that, in due course, lead on to the formation of carbon and hence of all heavier elements.[20]

Ends and Beginnings

And so, when our Universe was already over nine billion years old, a density fluctuation triggered, perhaps, by the explosion of a nearby supernova, led to gravitational infall in a region of our galaxy, still over-whelmingly consisting of primordial hydrogen and helium, but with a sprinkling of dust containing heavier elements formed by both r and s processes in previous generations of stars. Nearly all the matter in this cloud would end up in its central star, our Sun, and nearly all the rest would go to making the giant gas and ice planets, Jupiter, Saturn, Uranus and Neptune. Closer to the Sun, dust particles would begin to aggregate together. Some tiny fraction of this aggregated material would, after further adventures, fall to Earth in the form of meteorites, carrying with it evidence of its complex origins.[21] Most would aggregate to form the rocky planets and, too close to Jupiter to settle down and form a single body, the asteroid belt. Our own planet accreted over some tens of millions of years,[22] before collision with a Mars-size body, and redistribution of its materials, gave rise some four and a half billion years ago to the present Earth–Moon system.

Which is roughly where we came in.

Endnotes

Notes to Chapter 1 — The Age of the Earth — An Age-Old Question

1. One Mayan Long Count calendar cycle of 5125 years comes to an end around December 21, 2012, a fact to which some strange people, for some strange reason, attach great significance.
2. c. 460 BC–370 BC. Actually, it is difficult to know which ideas were his, and which came from his mentor Leucippus. More on this when we come to discuss atomic theory.
3. Now best remembered for having discovered the law of the spring (extension is proportional to force).
4. Actually, that's not quite how he put it. Prout convinced himself that all gases had a density that was an exact multiple of the density of hydrogen, and inferred that their components were formed by condensation from hydrogen. The conventional formulation, which I've followed here for simplicity, is what follows if one combines Prout's hypothesis with Avogadro's.
5. See Simon Winchester, *The Map that Changed the World*, Penguin (2002).
6. S.A. Bowring, I.S. Williams (1999), Priscoan (4.00–4.03 Gyr) orthogneisses from northwestern Canada, *Contributions to Mineralogy and Petrology*, **134**, 3; M. Hopkins, T.M. Harrison, C.E. Manning (2008), Low heat flow inferred from >4 Gyr zircons suggests Hadean plate boundary interactions, *Nature*, **456**, 493, and references

therein; O. Abramov, S.J. Mojzsis (2009), Microbial habitability of the Hadean Earth during the late heavy bombardment, *Nature*, **459**, 419.

7. For example, the world's largest deposits of iron ore were laid down between 2.5 and 1.8 billion years ago, when the buildup of oxygen from photosynthetic bacteria was converting the iron in the oceans from relatively soluble chemical forms to something more like rust.

8. We call him Kelvin for simplicity, although he was not granted this title until 1892.

9. Joe D. Burchfield, *Lord Kelvin and the Age of the Earth*, University of Chicago Press (1990) has a fuller account of this controversy, the geologists' responses, and the uses made of other data such as the dynamics of the Earth–Moon system.

10. Precambrian (culminating in Ediacaran; officially added 2004), Cambrian, Ordovician, Silurian, Devonian, Carboniferous (often divided into Mississippian and Pennsylvanian), Permian, Triassic, Jurassic, Cretaceous, Palaeocene, Eocene, Oligocene, Miocene, Pliocene, Pleistocene, Holocene.

11. Macmillan's Magazine, vol. 5 (March 5, 1862), pp. 288–293. From reprint in Popular Lectures and Addresses, vol. 1, 2nd edition, pp. 349–368, through http://www.archive.org/details/popularlec-turesa01kelvuoft (retrieved 25 Sept 2011).

12. Culminating in his 1897 address to the Victoria Institute, annotated and published in *Science*, (1899), **9**, 665 and 704.

13. You may have noticed that naturally occurring uranium is actually a mixture of *three* isotopes. Of these, uranium-238 and uranium-235 are ancient, while uranium-234 is present in trace amounts as an intermediate in the decay of uranium-238 to lead. The existence of isotopes, and the importance of this in understanding the decay sequences of radioactive materials, was not properly realised until 1913.

14. A. Bouvier, M. Wadhwa (2010), The age of the Solar System redefined by the oldest Pb–Pb age of a meteoritic inclusion, *Nature Geoscience*, **3**, 637.

15. G.B. Dalrymple (2001), The age of the Earth in the twentieth century: A problem (mostly) solved, in C.L.E. Lewis, S.J. Knell (eds.), *The Age of the Earth: from 4004 BC to AD 2002*, Geological Society of London Special Publication **190**, p. 205; Y. Amelin, A.N. Krot, I.D. Hutcheon, A.A. Ulyanov (2002), Lead Isotopic Ages of Chondrules and Calcium-Aluminum-Rich Inclusions, *Science*, **297**, 1678.

We can expect further marginal refinement of the accuracy of these measurements, taking into account the different fates of various elements in the early Solar System. Consider a radioisotope with a half life of the few million years. This isotope, and its stable decay product, may well tend to end up in different chemical environments. So if we find the decay product in an environment more appropriate to the parent, we can assume that it found its way there before that parent decayed. See G. Brennecka *et al.* (2010), ^{238}U/^{235}U Variations in Meteorites: Extant ^{247}Cm and Implications for Pb-Pb Dating, *Science*, **327**, 449 (extant here means extant in the early Solar System); T.W. Dahl, D.J. Stevenson (2010), Turbulent mixing of metal and silicate during planet accretion — And interpretation of the Hf–W chronometer, *Earth and Planetary Science Letters*, **295**, 177.

16. A.N. Halliday, B.J. Wood (2009), How Did Earth Accrete? *Science*, **325**, 44; J.F. Rudge, T. Kleine, B. Bourdon (2010), Broad bounds on Earth's accretion and core formation constrained by geochemical models, *Nature Geoscience*, **3**, 439.

17. B.B. Boltwood (1907), On the Ultimate Disintegration Products of the Radioactive Elements. Part II. The Disintegration Products of Uranium, *American Journal of Science*, **23**, 77; L. Badash (1968), Rutherford, Boltwood, and the Age of the Earth: The Origin of Radioactive Dating Techniques, *Proceedings of the American Philosophical Society*, **112**, 157.

18. For more on Holmes, see his biography; Cherry Lewis, *The Dating Game*, Cambridge University Press (2000).

19. Although two notable geologists did in fact make the perfectly valid point that Kelvin was making a potentially unwarranted assumption when he claimed that all sources of heat and energy had been accounted for. See James Croll (1877), On the probable origin and age of the Sun, *Quart. J. Sci.*, **7**, 307, and especially T.C. Chamberlin's 1899 response to Kelvin's 1897 Victoria Institute address, *Science*, **9**, 889 and **10**, 11, in which Chamberlin speculates on changes by which matter at the pressures and temperatures to be found at the centre of the Sun might release large amounts of energy through hitherto unknown interactions between the constituent atoms. Chamberlin was far ahead of his time in this suggestion, as in his suggestion that cometary impact on the early Earth could have been responsible for delivering the water in the oceans.

Notes to Chapter 2 — Atoms Old and New

1. Lucretius, *On the Nature of the Universe* (ca. 55 BC), my translation.
2. This statement is an example of a scientifically useful tautology. There are, of course, processes (the processes of radioactive decay, and those of nuclear fusion that we discuss in the final Chapter) where the number of atoms of each kind is not conserved because one element is transformed into another. We simply decide to call these physical processes, so that our statement remains true by definition. Nonetheless, it is useful, because it is usually pretty obvious whether a process should be called "chemical" or "physical", on other grounds, such as whether or not it involves the formation of new bonds between atoms.
3. In present-day notation,

$$C + O_2 \rightarrow CO_2$$

and

$$CaCO_3 \rightarrow CaO + CO_2$$

4. This is not quite true, as we saw when discussing radioactive dating. Most elements are a mixture of atoms of slightly different mass but almost identical chemical properties. The relative atomic masses of the elements as they occur in nature are an average of the masses of these *isotopes*.
5. So we can write the reactions as

$$H_2 + Cl_2 \rightarrow 2HCl$$

and

$$2H_2 + O_2 \rightarrow 2H_2O$$

6. The principle used here is that the force exerted by a magnetic field on a moving charge depends on the size of the charge and how fast it's moving. If there is no force on an object, it carries on moving in the same straight line. The "centrifugal force" required to make an object turn increases with the tightness of the turn and the speed and mass of the object; this is why a car skids if you try to take a curve too tight or too fast. It turns out that the experiment contains enough information to let us calculate the ratio of the mass (which determines

the inertia, or tendency to keep going straight on) to the charge (which determines the force making the particle move round through a curve), but not enough to find them out separately.

7. Strictly speaking, once again, the ratio of mass to charge, but the charge of the electron had by this time been independently determined by balancing the electrostatic force on a charged oil drop against its weight.

8. Not an exact whole number, for two reasons. The neutron, and the proton together with the electron that balances its charge, both weigh slightly more than one atomic mass unit. And the mass of an atom is slightly less than what would be calculated by adding up the masses of its components, because of the energy given out in its formation.

Notes to Chapter 3 — The Banker Who Lost His Head

1. The confusing names and dates result from overlap between specific wars between France and Britain in Europe and in North America, and more general conflict in Europe.

2. Here, and throughout my account of the French Revolution, I am well aware of greatly simplifying complex events. For example, the Jacobin Terror was followed in its turn by a "White Terror", inflicted on those associated with the most radical of the revolutionaries.

3. These days, of course, we ideally use double-blind experiments, in which neither the subject, nor the experimenter, knows which drugs or procedures are genuine, and which are controls.

4. Magnet therapy, Skeptics Dictionary, http://www.skepdic.com/magnetic.html (retrieved 22 Aug 2011).

5. M. Shermer (Nov. 2002), Mesmerized by Magnetism, *Scientific American*, **287(5)**, 41.

6. R.P. Multhauf (1962), On the Use of the Balance in Chemistry, *Proc. Amer. Phil. Soc.*, **106**, 210.

7. *Elements of Chemistry*, Dover edition (1965), author's preface, p. xix. I fear much the same could be said about the present-day introductory curriculum in chemistry, where students are immediately exposed to concepts derived from quantum mechanics and thermodynamics, without any basis in experience to give these concepts meaning.

Notes to Chapter 4 — From Particles to Molecules, with A Note On Homoeopathy

1. For more on Dalton, see Frank Greenaway, *John Dalton and the Atom*, Heinemann/Cornell University Press (1966), and Arnold Thackray, *John Dalton: Critical Assessments of His Life and Science*, Harvard University Press (1972); I have followed these closely regarding Dalton's thinking about atoms. For Avogadro and the history of his hypothesis, see Mario Morselli, *Amedeo Avogadro: A Scientific Biography (Chemists and Chemistry series)* Springer (1984).

2. Boyle's Law states that gas pressure, at fixed temperature, is inversely proportional to volume. So if, as here, we expand the volume of the nitrogen by a factor of 5/4, the pressure changes by a factor of 4/5.

3. For simplicity, I use such familiar modern terms as *atom, element, compound* and *molecule* except when directly quoting Dalton and his contemporaries, albeit at the risk of suggesting that the distinctions which are clear to us now were equally clear to them, then. I also refer to "nitrogen" where Dalton, following Lavoisier, spoke of "azote", and use the modern symbols for the different elements, and the modern conventions for writing chemical formulas.

4. When Avogadro used the term "integral molecule", he was referring to a molecule of a compound. But a molecule of an element was simply called a "molecule" without further qualification. Both "molecules" and "integral molecules" were made up of "elementary molecules", corresponding to what we call atoms.

5. I refer to valence bond theory.

6. Crystal field theory, now in fact replaced by the rather more general ligand field theory.

7. Pedantically speaking, relative atomic masses. In this section, I have used very approximate atomic weights, and compared them to a normal hydrogen atom. Present-day atomic weights, some of which have been measured to better than one part in 10 million, are expressed on a very slightly different scale where the most common isotope of carbon, ^{12}C, is equal to exactly 12. This is for reasons of experimental convenience, and makes no difference to the discussion here. I have also used the correct values for relative weights, and chosen examples for their familiarity and convenience, whether or not these happen to correspond to the information actually available at the time.

8. In *Murder in the Cathedral*, his hero, Thomas á Becket, describes it as the greatest treason.

9. The rule works because each atom has six *degrees of freedom*, corresponding to how far it is displaced from its position of greatest stability, and how quickly it is moving, in each of the *x*, *y*, and *z* directions. This simple explanation assumes that the energy of an atom can vary continuously. Actually, the possible energies are restricted by the laws of quantum mechanics, but at high temperatures the difference between these possible energies is small compared to the total amount of energy available.

10. We can think of a molecule of hydrogen, for example, spinning around an axis at right angles to its length, like a rotating dumb-bell.

11. Pressure depends on the number of impacts the molecules make against the walls, and the amount of momentum transmitted by each impact. The number of impacts increases in direct proportion to the speed of the molecules. The momentum transmitted by each impact depends on the momentum of the individual molecules, which is simply mass per molecule times speed. (Pedants may prefer the word "velocity", since what counts is the component of the speed at right angles to the wall.) So pressure is proportional to molecular mass times the square of the molecular speed. But this in turn is directly proportional to molecular kinetic energy.

12. *Nuevo Cimento*, **7**(1858), 321–366; English translation *Sketch of a Course of Chemical Philosophy*, Alembic club reprint No. 18, Edinburgh, 1910, authorised by Cannizzaro a few days before his death.

Notes to Chapter 5 — The Discovery of the Noble Gases — What's so New About Neon?

1. Best remembered by chemists as a chemist, and by physicists as a physicist. Among his other achievements, he was the first to measure Newton's universal gravitational constant, from the strength of attraction between lead weights.

2. And not only on this occasion. Cavendish also isolated two other elements, oxygen (from decomposition of mercuric oxide), and hydrogen (from the reaction between sulphuric acid and metallic iron), and even reacted them together to make water, but continued to misinterpret his results in terms of the phlogiston theory.

3. Why do the lines appear bright in a street lamp and dark in the Sun's spectrum? The dark lines are caused by atoms in their lowest energy state absorbing light emitted by the very hot gas underneath, while the bright emission lines are due to atoms that have acquired excess energy, in the electric discharge or in a flame, and give up that energy in the form of light.

4. Named after the same Henry Cavendish.

5. More precisely, relative masses.

6. Strictly speaking, Avogadro's hypothesis only applies to an "ideal gas", in which the actual volume of the gas molecules, and the forces between them, can be ignored. However, real gases tend towards ideal behaviour at low pressures. and Ramsay and Rayleigh, just as we would today, used data collected over a range of pressures to correct for non-ideality.

7. The usual way of doing this is to measure the density of the gas, and apply Avogadro's hypothesis. In their first experiments, Ramsay and Rayleigh did not have enough material to do this, so they measured the rate at which the gas under pressure escaped through a porous container. The heavier the molecules of a gas, the more slowly they move and escape.

8. 16th and 18th August, 1894; reported in *Chemical News*, 24 August, 1894, p. 87.

9. Argon has a mass around 40. N_3 would have a mass of 42. A mass of 40, if we arrange elements on that basis, puts argon in between two highly reactive metals, potassium and calcium. This is clearly unacceptable in view of its properties, and Mendeleev (On Argon, *Chemical News*, July 12, 1895, p. 14) rejected placing it in its correct place just before potassium because he regarded the existence of a new group of elements as "scarcely admissible".

10. Although Svante Arrhenius's suggestions to this effect were initially controversial, because of the far from straightforward behaviour of strong electrolyte solutions.

Notes to Chapter 6 — Science, War, and Morality; The Tragedy of Fritz Haber

1. Of the many available accounts of Haber's life and work, two are notable for the personal connections between their authors and the events they describe. *The Poisonous Cloud: Chemical Warfare in the First World War*, Clarendon Press (1986), a detailed history of the

topic, was written by Ludwig Fritz (Lutz) Haber, his son. *Einstein's German World*, Princeton University Press (1999; earlier partial version 1996), which contains a long essay on Einstein and Haber that I have relied on extensively in this chapter, is by the historian Fritz Stern, Haber's godson and son of his physician. That both these accounts were so long in coming to publication is to me evidence of the ongoing emotional impact of the events they describe. The best overall account is *Fritz Haber, Chemist, Nobel Laureate, German, Jew*, by Dietrich Stolzenberg (English version Chemical Heritage Press, 2004). *Between Genius and Genocide*, by Daniel Charles (Jonathan Cape, 2005) is more journalistic and melodramatic, but contains useful additional detail.

2. Quoted by Tobias Brinkmann (2007) in *Central European History*, **40**, 348; review of Uffa Jensen, *Gebildete Doppelgänger. Bürgerliche Juden und Protestanten im 19.Jahrhundert*, Vandenhoeck & Ruprecht, 2005.

3. Using standard chemical notation we have, for the direct synthesis of ammonia,

$$N_2 + 3H_2 \rightarrow 2NH_3$$

For burning ammonia in air,

$$4NH_3 + 3O_2 \rightarrow 2N_2 + 6H_2O$$

For the catalysed oxidation to NO and subsequent conversion to nitric acid,

$$4NH_3 + 5O_2 \rightarrow 4NO + 6H_2O$$

$$2NO + O_2 \rightarrow 2NO_2$$

$$4NO_2 + O_2 + 2H_2O \rightarrow 4HNO_3$$

and for neutralisation with ammonia,

$$4HNO_3 + 4NH_3 \rightarrow 4NH_4NO_3$$

4. On heating, $\qquad CaCO_3 \rightarrow CaO + CO_2$

   In electric arc furnace, $\qquad CaO + 3C \rightarrow CaC_2 + CO$

   and, on introduction of nitrogen, $\quad CaC_2 + N_2 \rightarrow CaCN_2$ (calcium cyanamide) $+ C$

   Reaction with steam then gives $\qquad CaCN_2 + 3H_2O \rightarrow CaCO_3 + 2NH_3$

5. In fact things are even more complicated. The reason the decomposition did not go to completion, is that at very low concentrations of ammonia it is quite slow. Equilibrium is not reached over copper because copper is not a catalyst for the reaction. It is only in the presence of a suitable catalyst that this reaction approaches equilibrium. So to the extent that they were inspired by Ramsay's result (and they were), the scientists involved in the next stage of the work were doing the right thing for the wrong reason. Science is like that.

6. The only stronger bond is that in carbon monoxide, CO, which is about 14% stronger. N_2 and CO are both held together by *triple bonds*, which involve the sharing of three pairs of electrons. H_2 is held together by a single bond, but this bond is one of the strongest single bonds. For more about bonding, see Chapter 12.

7. This is the absolute temperature, or temperature above absolute zero, and on the "Kelvin scale" used in scientific work these days is equal to the temperature Celsius + 273.15°. In this particular case, as discussed below, the bond breaking takes place on the surface of the catalyst rather than within the bulk of the gas, but the same principle applies.

8. Strictly speaking, we are using Avogadro's hypothesis, which says that under the same conditions of temperature and pressure, different gases contain the same number of molecules, together with Boyle's Law, which tells us how the volume decreases when we increase the pressure, and Dalton's Law of partial pressures, which says that the total pressure in a mixture of gases is the sum of the pressures that each gas would exert on its own.

9. It is still the case that the release of heat favours a reaction at any temperature, but the higher the temperature, the smaller this effect, so that the reaction becomes less favourable. This is because the amount of entropy associated with a given amount of heat is inversely proportional to temperature. Hence one alternative formulation of the Second Law: heat can only pass spontaneously from a hotter body to a colder body.

10. W. Nernst (research by F. Jost) (1907), Über das Ammoniakgleichgewicht (On the ammonia equilibrium), 14th Conference of the German Bunsen Society for General Physical Chemistry, *Z. Elektrochem.*, **13**, 521.

11. Ammonia is a gas at ordinary temperatures and pressures, but at 200 atm it remains liquid up to its *critical temperature* of 132°C.

12. This account closely follows that given by Robert Le Rossignol (1928), Zur Geschichte der Herstellung des synthetischen Ammoniaks (On the history of the production of synthetic ammonia), *Naturwissenschaften*, **16**, 1070.
13. *Nature* **109**, 40 (1922).
14. Trevor Davies in *Oxford Dictionary of National Biography*, (retrieved 23 Aug 2011). Lamb actually spent the war in Ireland, working on weather predictions for the first transatlantic civilian flights.
15. For his heat theorem, which dated back to 1906. There is more to this long delay than meets the eye; see Chapter 12.
16. *Guardian*, 30 September 2010.
17. Eventually, of course, such niceties would be set aside.
18. Bryan Reuben, private communication, based on conversations with Ludwig Spiro, then 98, son of Haber's cousin Ernst, spring 2010.
19. Weizmann Institute (Rehovot) archives; Haber to Weizmann Sept. 1933, 5 Oct. 1933, 6 Jan. 1934; Weizmann to Haber 2 Oct., 20 Oct. 1933. Note, however, that Haber did find the strength to visit Cambridge after his letter of 6 January, and that there is no clear consensus among historians as to his ultimate intentions.
20. Global Security Newswire, 5 July 2011, through http://gsn.nti.org/gsn/nw_20110705_3042.php (accessed 23 Aug 2011).
21. Sir William Ehrman, the Foreign Office's director of international security at the time, evidence to UK Iraq War Enquiry, 24th November 2009.
22. See P. Iyngaran, D.C. Madden, S.J. Jenkins, D.A. King (2011), *Surface Chemistry Special Feature:* Hydrogenation of N over Fe{111}, *PNAS*, **108**, 925.
23. See J.A. Pool, E. Lobkovsky, P.J. Chirik (2004), Hydrogenation and cleavage of dinitrogen to ammonia with a zirconium complex, *Nature*, **427**, 527; D.J. Knobloch, E. Lobkovsky, P.J. Chirik (2009), Dinitrogen cleavage and functionalization by carbon monoxide promoted by a hafnium complex, *Nature Chemistry*, **2**, 30.
24. *The state of food insecurity in the world*, FAO, http://www.fao.org/publications/sofi/en/ (accessed 23 Aug 2011).
25. Special 30th Anniversary Lecture, The Norwegian Nobel Institute, Oslo, September 8, 2000.

Notes to Chapter 7 — The Ozone Hole Story — A Mystery with Three Suspects

1. The lowest temperature possible, at which, in classical physics, all molecular motion would cease, although quantum mechanics requires systems to retain an irreducible amount of *zero point energy*.
2. According to the equations

$$O_2 + \text{light energy} \rightarrow 2\,O$$

and

$$O + O_2 \rightarrow O_3$$

Actually, things are slightly more complicated. O_3 formed in this way will have enough energy to tear itself apart again, unless it happens to collide with another molecule that carries away its surplus energy, so this last equation is often written

$$O + O_2 + X \rightarrow O_3 + X$$

For the eventual destruction of the ozone, we have simply

$$O_3 + O \rightarrow 2\,O_2$$

and a third body is not needed because the surplus energy is carried away as kinetic energy by the separating O_2 molecules.
3. Ozone itself is split up by UV light into O_2 and O atoms, but since most of these O atoms find a molecule of O_2 and regenerate ozone, we can ignore this complication.
4. NASA *Ozone Facts: History of the Ozone Hole*, http://ozonewatch.gsfc. nasa.gov/facts/history.html, (retrieved 23 Aug 2011).
5. J. B. Kerr, C. T. McElroy (1993), Evidence for Large Upward Trends of Ultraviolet-B Radiation Linked to Ozone Depletion, *Science*, **262**, 1032.
6. J.F.Abarca, C.C. Casiccia (2002), Skin cancer and ultraviolet-B radiation under the Antarctic ozone hole: Southern Chile, 1987–2000, *Photodermatol. Photoimmunol. Photomed.*, **18**, 294.
7. Or, as it was known until 1961, Stockholm Högskola.
8. Involving the hydroperoxide radical as an intermediate:

$$OH + O_3 \rightarrow HO_2 + O_2$$
$$HO_2 + O_3 \rightarrow OH + 2O_2$$

9. The relevant reactions are

$$NO + O_3 \rightarrow NO_2 + O_2$$

$$NO + O + X \rightarrow NO_2 + X$$

which is cancelled out by the reactions

$$NO_2 + \text{light energy} \rightarrow NO + O$$

$$O + O_2 + X \rightarrow O_3 + X$$

except for the 5% NO_2 that is removed by

$$NO_2 + O \rightarrow NO + O_2$$

Smog formation at lower levels by way of intermediate NO_2 takes place through a more complicated reaction sequence, with overall effect

$$2\,NO + O_2 \rightarrow 2\,NO_2$$

10. $O_3 + Cl \rightarrow O_2 + ClO$, followed by $ClO + O \rightarrow Cl + O_2$

11.
$$NO + Cl \rightarrow ClNO$$

$$ClNO + O_2 \rightarrow ClNO_3$$

$$HCl + ClNO_3 \rightarrow HNO_3 + Cl_2$$

$$Cl_2 + \text{light energy} \rightarrow 2Cl$$

12. Unknown in Britain, but quite common in some US cities. People are advised to stay indoors as much as possible, and not open their windows.

13. Why this difference? Because near the surface, molecular oxygen is abundant, but ozone and, especially, oxygen atoms are relatively rare. In the stratosphere, the amounts of "odd-numbered oxygen" (atomic oxygen, together with ozone) are relatively far higher, and the concentration of O_2 far lower, than at the surface.

14. Scientific Assessment Panel Of The Montreal Protocol On Substances That Deplete The Ozone Layer, *Scientific Assessment of Ozone Depletion: 2010*, http://www.wmo.int/pages/mediacentre/press_releases/documents/898_ExecutiveSummary.pdf, (retrieved 23 Aug 2011).

15. G. L. Manney *et al.* (2011), Unprecedented Arctic Ozone Loss in 2011, *Nature*, doi:10.1038/nature 10556, published online 02 October.

Notes to Chapter 8 — Rain Gauge, Thermometer, Calendar, Warning

1. P. Zhang, *et al.* (2008), A Test of Climate, Sun, and Culture Relationships from an 1810-Year Chinese Cave Record, *Science,* **322**, 940.
2. Specifically, zero point energy, which is proportional to vibrational frequency, and decreases with mass; see Chapter 11 for more on this.
3. X. Wang *et al.* (2010), Climate, Desertification, and the Rise and Collapse of China's Historical Dynasties, *Hum. Ecol.,* **38**, 157.
4. Z. Zhang *et al.* (2010), Periodic climate cooling enhanced natural disasters and wars in China during AD 10–1900, *Proc. R. Soc. B,* **277**, 3745.
5. This section title is borrowed with permission from *The Tree Rings' Tale,* John Fleck, University of New Mexico Press, 2009.
6. Review: P.B. deMenocal (2001), Cultural Responses to Climate Change During the Late Holocene, *Science,* **292**, 667.
7. H.M. Cullen *et al.* (2000), Climate change and the collapse of the Akkadian empire: Evidence from the deep sea, *Geology,* **28**, 379.
8. L.C. Peterson, G.H. Haug (2005), Climate and the collapse of Maya civilisation: a series of multi-year droughts helped to doom an ancient culture, *American Scientist,* **93**, 322.
9. M. Medina-Elizalde *et al.* (2010), High resolution stalagmite climate record from the Yucatán Peninsula spanning the Maya terminal classic period, *Earth and Planetary Science Letters,* **298**, 255.
10. U. Büntgen *et al.* (2011), 2500 Years of European Climate Variability and Human Susceptibility, *Science,* **331**, 578.
11. I.J. Orland *et al.* (2009), Climate deterioration in the Eastern Mediterranean as revealed by ion microprobe analysis of a speleothem that grew from 2.2 to 0.9 ka in Soreq Cave, Israel, *Quaternary Research,* **71**, 27.
12. C.J. Caseldine, C. Turney (2010), The bigger picture: towards integrating palaeoclimate and environmental data with a history of societal change, *J. Quaternary Science,* **25**, 88.

Notes to Chapter 9 — Making Metal

1. Using the obvious transcriptions from Greek to Latin alphabet; from Liddell, Scott, and Jones, Clarendon Press. 1940.
2. This is what is called specular, or mirror-like, reflection. It is a little different in detail from the absorption of light discussed in Chapter 13.

3. So we have the paradox of a self-defeating prediction, related to the computational uncertainty discussed in Chapter 10.

4. Perhaps tactlessly, since the gold was, after all, on the other side of the border with Canada.

5. $4Au + 8KCN + O_2 + 2H_2O \rightarrow 4KAu(CN)_2 + 4KOH$

6. For copper oxide ore, the relevant reactions are

$$2C + O_2 \rightarrow 2CO$$

$$CO + CuO \rightarrow CO_2 + Cu$$

$$CO_2 + C \rightarrow 2CO$$

Note that I did not say "is stronger than". The formation of CO and CO_2 is entropy-favoured (see Chapter 6) and can take place at high enough temperature even if the reaction absorbs energy.

7. $$CuS + O_2 \rightarrow Cu + SO_2$$

$$2SO_2 + 2H_2O + O_2 \rightarrow 2H_2SO_4$$

8. $CuSO_4 + Fe \rightarrow FeSO_4 + Cu$,
   or, ignoring the sulphate spectator ion,

$$Cu^{2+} + Fe \rightarrow Cu + Fe^{2+}$$

9. See e.g. K.H. Debus (1990), Mining With Microbes, *Technology Review*, **93**, 50.

10. R. Maddin, J.D. Muhly, T. Stech (1999), Early metalworking at Çayönü, in A. Hauptmann *et al.*, eds., *Der Anschnitt*, **9** (The Beginnings of Metallurgy), Deutsches Beergbau Museum, p. 37.

11. M. Radivojević *et al.* (2010), On the origins of extractive metallurgy: new evidence from Europe, *Journal of Archaeological Science*, **37**, 2775.

12. Remember that lead is the final product of radioactive disintegration. Minerals originally containing different amounts of the long-lived radioactive parents (^{232}Th, ^{235}U, ^{238}U) will eventually give rise to different isotopes of lead. For this reason, the isotopic composition of lead is particularly sensitive to its geological origin and geographical provenance.

13. The Latin letters U and Y both derive from the same Greek letter, which changed its pronunciation in pre-classical times.

14. See K.A. Yener *et al.*, Analysis of Metalliferous Residues [etc] from Kestel Tin Mine and the Tin Processing Side of Göltepe, Turkey, in

P.T. Craddock, J. Lang, eds., *Mining and Metal Production Through The Ages*, British Museum Press, 2003; *The Problem of Early Tin*, A. Giumlia-Mair, and F. Lo Schiavo, eds., BAR/Archaeopress, 2003.

15. N. Nezafati, E. Pernicka, M. Momenzadeh (2006), Ancient Tin, Old Question & A New Answer, http://www.cais-soas.com/CAIS/Science/tin.htm, (retrieved 23 Aug 2011); abstracted from *Antiquity*, **80** (308).

16. Congo's Riches, Looted by Renegade Troops, Lydia Polgreen, New York Times, November 15, 2008.

17. D.M. Jacobson (2000), Corinthian bronze and the gold of the alchemists, *Gold Bulletin*, **33**, 60.

18. G.A Rendsburg (1982), Semitic PRZL/BRZL/BRDL, *Scripta Mediterranea* **3**, 54, available via http://jewishstudies.rutgers.edu/component/docman/doc_view/50-semitic-przlbrzlbrdl-iron, (retrieved 23 Aug 2011); BRZL incorporates *Bar*.

19. Although often referred to as an "alloy" between iron and carbon, cementite, with the formula Fe_3C, is a distinct compound that contains about 7% carbon by weight.

20. P. T. Craddock, Cast Iron, Fined Iron, Crucible Steel: Liquid Iron in the Ancient World, in P.T. Craddock, J. Lang eds., *Mining and metal production through the ages*, British Museum Press, 2003, p. 231.

21. For a fuller discussion of the ceremonial Japanese sword, see M.R. Notis (2000), The history of the metallographic study of the Japanese sword, *Materials Characterization*, **45**, 253.

22. See K.R. Maxwell-Hyslop (1972), The metals *amūtu* and *aši'u* in the Kültepe texts, *Anatolian Studies*, **22**, 159; J. D. Muhly *et al.* (1985), Iron in Anatolia and the nature of the Hittite iron industry, *Anatolian Studies*, **35**, 67; Ü. Yalçin (1999), Early iron metallurgy in Anatolia, *Anatolian Studies*, **49**, 177.

23. T. Stech-Wheeler *et al.* (1981), Iron at Taanach and Early Iron Metallurgy in the Eastern Mediterranean, *American Journal of Archaeology*, **85**, 245.

24. R. Tewari (2003), The origins of iron working in India: New evidence from the Central Ganga Plain and the Eastern Vindhyas, *Antiquity*, **77**, 536, and references therein.

25. S.B. Alpern (2005), Did They or Didn't They Invent It? Iron in Sub-Saharan Africa, *History in Africa*, **32**, 41.

26. I Samuel 13:19–20. The books of Samuel show evidence of contributions from anti-monarchists, supporters of the House of Saul, and supporters of the House of David. All of these shared an

interest in describing the playing field as tilted in favour of the Philistines, in order to excuse their own failings or to exaggerate their successes.

27. More specifically, perhaps, the Judaeans, but this would be true of both Kingdoms. The relevant verses are Deuteronomy 8:9 for iron ore extraction, and 27:5 (and also Joshua 8:31) for the ban on iron tools when building an altar. Some fundamentalists and extreme traditionalists will reject my suggested dating as blasphemous; too bad.

I take no position here on the historicity of the Books of Samuel, beyond noting that the incursions of the Philistines, and the existence by the 9th century BC of a dynasty known as the House of David, are fully attested elsewhere.

28. R. Bentley, T.G. Chasteen (2002), Arsenic Curiosa and Humanity, *Chemical Educator,* **7**, 51.

29. F. Mari, A. Polettini, D. Lippi, E. Bertol (2006), The mysterious death of Francesco I de' Medici and Bianca Cappello: an arsenic murder? *BMJ,* **333**, 1299.

30. In this, arsenic is converted to its hydride, AsH_3, by metallic zinc in the presence of acid, and leaves a characteristic black deposit of free arsenic if the flame produced by burning the hydrogen generated at the same time is allowed to play against a cold ceramic surface.

31. J.T. Hindmarsh, J. Savory (2008), The Death of Napoleon, Cancer or Arsenic? *Clinical Chemistry,* **54**, 2092.

32. S. Fendorf, H.A. Michael, A. van Geen (2010), Spatial and Temporal Variations of Groundwater Arsenic in South and Southeast Asia, *Science,* **328**, 1123.

33. F. Wolfe-Simon *et al.* (2011), A Bacterium That Can Grow by Using Arsenic Instead of Phosphorus, *Science,* **332**, 1163.

34. See *Science,* **332** (3 June 2011) pp 1136, 1149, 1163 (multiple authors).

35. According to *The Poisonous Cloud: Chemical Warfare in the First World War,* L.F. Haber, Clarendon Press (1986). See also Chapter 6 for a discussion of the limitations of such chemical weapons.

36. Mercury forms two series of compounds, mercuric compounds in which mercury carries an average formal charge of +2, and mercurous compounds, where this charge is +1. Actually, the mercurous compounds contain Hg_2^{2+} units, in which the two mercury atoms, as well as carrying a positive charge, are chemically bonded to each other.

37. There are as many as three additional liquid layers; silicate slag, a sulphide-rich layer, and a layer containing arsenic, copper, and iron.

These may be worth working up independently, or adding to later batches of ore.

38. S. Hong, J.-P. Candelone, C.C. Patterson, C.F. Boutron (1994), Greenland Ice Evidence of Hemispheric Lead Pollution Two Millennia Ago by Greek and Roman Civilizations, *Science*, **265**, 1841.

39. For the myth, and a scathing refutation, see S.C. Gilfillan (1965), Lead Poisoning and the Fall of Rome, *Journal of Occupational Medicine*, **7**, 53; Jerome O. Nriagu, *Lead and Lead Poisoning in Antiquity*, Wiley, 1983; J.O. Nriagu (1983), Saturnine Gout Among Roman Aristocrats: Did Lead Poisoning Contribute to the Fall of the Empire? *New England Journal of Medicine*, **308**, 660; J. Scarborough (1984), The Myth of Lead Poisoning Among the Romans: An Essay Review, *Journal of the History of Medicine*, **39**, 469.

40. J. Eisinger (1982), Lead and wine: Eberhard Gockel and the *colica Pictonum*, *Medical History*, **26**, 279.

41. Lead poisoning kills 400 more Nigerian children, *Reuters*, 7 March 2011, http://www.reuters.com/article/2011/03/07/us-nigeria-poisoning-idUSTRE7264IC20110307 , (retrieved 23 Aug 2011).

42. Actually, it's an electrochemical effect. Zinc holds onto its electrons less tightly than iron, so it builds up a bit of a positive voltage and the iron becomes slightly negative. This makes it more difficult for the iron to react, because that involves it separating from the bulk metal as a positive ion. But meanwhile, the zinc can't react either because of the protective coating.

43. Sulphuric acid will react with zinc oxide, sulphide, or carbonate ores to give zinc sulphate in solution:

$$ZnO + H_2SO_4 \rightarrow ZnSO_4 + H_2O$$

$$ZnS + H_2SO_4 \rightarrow ZnSO_4 + H_2S$$

$$ZnCO_3 + H_2SO_4 \rightarrow ZnSO_4 + H_2O + CO_2$$

Zinc sulphate solution actually contains Zn^{2+} and SO_4^{2-} ions, The electrochemical reactions are

$$Zn^{2+}(\text{in solution}) + 2e^- \rightarrow Zn(\text{deposited solid})$$

$$2H_2O \rightarrow 4H^+ + O_2(g) + 4e^-$$

H^+ is the dissolved hydrogen ion, which is strongly attached to water, and e^- is an electron gained or lost at an electrode. Since the number of electrons lost at one electrode must balance the number gained at the other, two atoms of zinc are deposited for every molecule of O_2 produced. So starting with zinc oxide, the overall reaction is

$$2ZnO \rightarrow 2Zn + O_2$$

The H^+ formed in solution when oxygen is generated, together with the sulphate spectator ion, constitute regenerated sulphuric acid. Thermodynamically, it should be easier to reduce the solution to hydrogen than to extract the zinc. However, doing this would mean making hydrogen atoms as an intermediate step, and zinc is a metal that does not make surface bonds to hydrogen (contrast the discussion of catalytic metals in Chapter 6).

44. The relevant equations are
    Production of quicklime: $CaCO_3 \rightarrow CaO + CO_2$
    Reaction and dissolution: $CaO + H_2O \rightarrow Ca(OH)_2$ = (in solution)
    $Ca^{2+} + 2OH^-$
    Precipitation: $2OH^- + Mg^{2+} \rightarrow Mg(OH)_2$ (solid)
    Neutralisation: $Mg(OH)_2$ (solid) + 2HCl (solution) →
    $MgCl_2$ (solution) + $2H_2O$
    Followed by evaporation of water, melting the $MgCl_2$,
    and electrolysis: $Mg^{2+} + 2e^- \rightarrow Mg$ (liquid) (reducing electrode) and
    $2Cl^- \rightarrow Cl_2$ (gas) + $2e^-$ (oxidising electrode)

45. The key reaction is $2MgO + Si \rightleftharpoons 2Mg(vapour) + SiO_2$
    An ingenious refinement is to use a mixture of calcium oxide and magnesium oxide, prepared from dolomite, which can simply be quarried rather than laboriously extracted from seawater. For practical reasons, an iron-silicon alloy is used rather than silicon itself. This ferrosilicon is prepared by heating together scrap iron, sand, and coke, a reaction which again is heat-absorbing, but favoured by entropy because it generates carbon dioxide gas. The full reaction sequence then is

$$Fe + SiO_2 + C \rightarrow FeSi + CO_2$$

$$CaCO_3.MgCO_3 \text{ (dolomite)} \rightarrow CaO.MgO$$
$$\text{(intimately mixed oxides)} + 2CO_2$$

$$2CaO.MgO + FeSi \rightleftharpoons 2Mg + Ca_2SiO_4 + Fe$$

The use of reduced pressure illustrates Le Châtelier's principle (again; see Chapter 6). The calcium silicate produced is incorporated into cement.

46. $Na^+(melt) + e^- \rightarrow Na(liquid, at these temperatures)$

$$2Cl^- \rightarrow Cl_2(gas) + 2e^-$$

Two Na are produced for each Cl_2. Of course, the cell is carefully designed to keep the sodium and chlorine well separated.

47. At the mercury electrode, in the amalgam process,

$$Na^+(solution) + e^- \rightarrow Na(amalgam)$$

As before, $2Cl^- \rightarrow Cl_2(gas) + 2e^-$

In a separate compartment, the sodium amalgam becomes the oxidising electrode, reversing the reaction, and we have

$$Na(amalgam) \rightarrow Na^+(solution) + e^-$$

while at the reducing electrode we have

$$2H_2O + 2e^- \rightarrow H_2(gas) + 2\ OH^-$$

so that in this second compartment we have a buildup of Na^+ and OH^- ions, i.e. of sodium hydroxide.

In the diaphragm cell, we have the now familiar reaction

$$2Cl^- \rightarrow Cl_2(gas) + 2e^-$$

at the oxidising electrode, and the reduction of water

$$2H_2O + 2e^- \rightarrow H_2(gas) + 2\ OH^-$$

at the other. The overall effect is to "subtract" the elements of HCl from sodium chloride in water, leaving sodium hydroxide behind.

48. $Al(OH)_3 + 3HCl(aqueous) \rightarrow AlCl_3(aqueous) + 3H_2O$

Concentrating this solution gives $Al(H_2O)_6Cl_3$.

On heating,

$$2Al(H_2O)_6Cl_3 \rightarrow Al_2O_3 + 6HCl + 9H_2O$$

49. $2Al_2O_3 + 3CCl_4 \rightarrow 2Al_2Cl_6 + 3CO_2$

Why Al_2Cl_6? See main text.

50. Simplified equations:

$$Al^{3+} + 3e^- \rightarrow Al; \quad 2O^{2-} \rightarrow O_2 + 4e^-$$

However, the reducing electrodes (made of carbon) are oxidised at a rate of, roughly, one atom carbon for every atom aluminium formed, and the full chemistry of what is going on is still not completely understood.

51. As in the preparation of aluminium chloride, carbon reacts with chlorine and the carbon tetrachloride formed undergoes an exchange reaction with the metal oxide:

$$C + 2Cl_2 \rightarrow CCl_4$$

$$CCl_4 + TiO_2 \rightarrow TiCl_4 + CO_2$$

The titanium tetrachloride reacts with water to give titanium dioxide:

$$TiCl_4 + 2H_2O \rightarrow TiO_2 + 4HCl$$

Or with magnesium, which, being more reactive than titanium, can displace it:

$$TiCl_4 + 2Mg \rightarrow Ti + 2MgCl_2$$

52. Rare earth metals mine is key to US control over hi-tech future; *Guardian*, 26 December 2010.

53. Japan develops vehicle motor free of rare earths, http://www.spacemart.com/reports/Japan_develops_vehicle_motor_free_of_rare_earths_999.html, 30 Sept. 2010, (retrieved 23 Aug 2011).

54. http://www.terradaily.com/reports/Japan_to_offer_rare_earth_subsidies_for_rare_earths_999.html, 25 February 2011, (retrieved 23 Aug 2011).

Notes to Chapter 10 − In Praise of Uncertainty

1. Among other places, this idea is eloquently developed in *Caveman Logic: The Persistence of Primitive Thinking in a Modern World* (Prometheus Books, 2009), by the guitarist and psychology professor Hank Davis.

2. A.V. Bell, P. J. Richerson, R. McElreath (2009), Culture rather than genes provides greater scope for the evolution of large-scale human prosociality, *PNAS*, **106**, 17671.

3. As discussed in the next chapter, Max Planck introduced the idea of the *quantum*, or smallest amount of a physical quantity that could be generated, in order to explain why glowing objects do not emit copious amounts of X-rays.

4. See Frank Wiczek, *The Lightness of Being: Mass, Ether, and the Unification of Forces*, Basic Books (2010).
5. P.-M. Binder (2008), Philosophy of science: Theories of almost everything, *Nature*, **455**, 884.

Notes to Chapter 11 — Everything is Fuzzy

1. The angle of diffraction, θ, for wavelength λ and spacing d is given by $n\lambda = d \sin\theta$, where n is a whole number. For a wavelength just a quarter of the spacing, the angles of diffraction are 14.5, 30, and 48.6 degrees.
2. If indigo really counts as a separate colour. Some people think that Newton included this among the colours of the spectrum because he thought there should be seven of them, to match the seven separate notes of the octave.
3. See e.g. P.C.W. Davies and J.R. Brown, *The Ghost in the Atom: A Discussion of the Mysteries of Quantum Physics*, Cambridge University Press (1986), which includes interviews with Alain Aspect, John Bell, and David Bohm, among others, or the more recent *Quantum Physics: Illusion or Reality?* 2nd ed., Alastair Rae, Cambridge University Press (2004).
4. In the ideal case, *black body radiation*. A black body is one which can absorb or emit radiation with perfect efficiency at any frequency.
5. When the universe was still too hot for atoms to hold on to their outermost electrons, space was filled with an electrically conducting plasma that absorbed light and prevented it from spreading.
6. Wavelength (distance per cycle) × frequency (cycles per second) = speed of wave (distance per second).
7. Although it is now known that some rather exceptional processes involve two or even more quanta in a single step.
8. This direct relationship only applies to the most loosely held, and therefore fastest moving, electrons. We now know that electrons are held inside a solid with a wide overall range of binding energies, and by analysing the range of energies with which electrons are emitted, we can build up a picture of this spread of energy levels, a technique known as photoelectron spectroscopy.

9. Much of the personal information I use here about Bohr and his circle comes from Abraham Pais, *The Genius of Science: A Portrait Gallery*, Oxford University Press (2000); Gino Segré, *Faust in Copenhagen: A Struggle for the Soul of Physics*, Jonathan Cape/Viking (2007); S. Rozental (ed.), *Niels Bohr, his life and work as seen by his friends and colleagues*, North-Holland (1967).

10. Die Plancksche Theorie der Strahlung und die Theorie der spezifis-chen Wärme (Planck's theory of radiation and the theory of specific heat), *Ann. Physik*, **22**, 180 (1907).

11. Using the symbols now current, kinetic energy = $\frac{1}{2} \times h \times$ frequency $\times$ n, where n is a whole number, and in the most stable state of the atom, $n = 1$. n is what we now call the *principal quantum number* of an elec-tron in an atom. Why $\frac{1}{2}$? Perhaps because the kinetic energy is equal in size to half the potential energy, or perhaps because it leads to the elegant result that angular momentum = $nh/2\pi$, where n is a whole number and 2π is the ratio of the circumference of a circle to the radius, or perhaps just because it gave the right answer. From the outset, this "old quantum theory" was regarded as a ramshackle expedient while the crumbling structure of classical physics was being replaced.

12. *Phil. Mag.*, **26**, 1, 476, 857 (1913).

13. Bohr's model also explained why the processes that gave rise to absorption and emission of radiation did not contribute to specific heat, as they should have done in classical theory.

14. Young actually used a thin card to split a beam of light, but that does not change the principle.

15. Indeed, uncertainty is too mild a term. It suggests incomplete know-ledge of something that is actually present in Nature. The Copenhagen interpretation, strictly speaking, refers to *indeterminacy*, meaning that the properties of a system are not determined until we observe it.

16. For example, we can generate two electrons with their spins (see next chapter) pointing in opposite directions. If you try to measure the direction of the spin, it will either be up, or down, relative to the mag-netic field you use to measure it, whatever the direction of the field. The paradox is that when you have measured the spin of one electron, the other electron seems to know instantly the direction of the meas-uring field you applied to the first one. This even though information cannot possibly flow from one electron to the other faster than the

speed of light. For a fuller discussion, see the books referred to in Note 3 to this chapter.

17. The cat is in a box, together with a sealed phial of hydrogen cyanide, and an electrical circuit that will be triggered to break the phial if and only if a radioactive atom happens to decay during a time when the box is sealed. The time of the radioactive decay is itself quantum mechanically indeterminate until observed. Does the circuit count as an observer, or the phial, or the (possibly) unfortunate cat, or do all of these hover between two possible states until someone opens the box and has a look?

18. De Broglie's specific suggestion was that we could associate a wavelength with any kind of object, and that this same relationship between wavelength and momentum would hold good for matter, not just for light.

   For anything, $E = mc^2$. But for a photon, $E = hf$, where f is the frequency. Hence $mc^2 = hf$ and $mc = hf/c$. For light, $c = v$ (velocity) and, for anything, mass x velocity = momentum, so

   $$\text{Momentum} = hf/c.$$

   And for any travelling wave, f (number of cycles per second) x wavelength (distance travelled in each cycle) must equal velocity (distance travelled per second), so that wavelength = velocity/frequency for any travelling wave and for light, specifically,

   $$\text{Wavelength} = c/f.$$

   Putting these equations together, we get de Broglie's equation:

   $$\text{Wavelength x Momentum} = c/f \times hf/c = h$$

19. I use here the usual "Copenhagen" interpretation of the wave in terms of probability, but de Broglie himself was among those dissatisfied with this.

20. V.P.A. Lonij, C.E. Klauss, W.F Holmgren, A.D. Cronin (2010), Atom Diffraction Reveals the Impact of Atomic Core Electrons on Atom-Surface Potentials, *Physical Review Letters,* **105**, 233202.

21. From Georg Duckwitz, German naval attaché in Copenhagen, who in September 1943 also visited Sweden to make sure that Sweden would receive the refugees. Brave man.

22. *Nobel Lectures, Physics, 1922–1941,* Elsevier (1965).

Notes to Chapter 12 — Why Things Have Shapes

1. Or, to use their rather clumsy but pedantically correct name, relative atomic masses.
2. There are many excellent discussions of the history and structure of the Periodic Table, among them *The Periodic Table*, by Eric R. Scerri, Oxford University Press (2007) and *The Periodic Kingdom* by P.W. Atkins, Basic Books (1995). *Nature's Building Blocks*, by John Emsley, Oxford University Press (2001) gives a complete catalogue in layman's terms of the elements and some of their most interesting properties. *The Disappearing Spoon*, by Sam Kean, Little, Brown (2010) is a well-deserved bestseller, while Primo Levi's incomparable *Il Sistema Periodico*, 1975, published in English as *The Periodic Table* (Everyman, 1996), a collection of essays using the elements as a framework, was listed by the Royal Society as among the best science books ever.
3. But hydrogen, coming just before helium, is a special case.
4. H.G.J. Moseley (1913), The high-frequency spectra of the elements, *Phil. Mag.*, **26**, 1024; (1914), Part II, *ibid.*, **27**, 703; Atomic Models and X-Ray Spectra, *Nature*, **92**, 554.
5. In Bohr's formula, frequency was proportional to the square of the atomic number N, while Moseley found his frequencies proportional (in exactly the same ratio) to the square of $(N - 1)$. Moseley correctly inferred that this difference was due to the other electrons in what he called the ring closest to the nucleus. We now know that this ring, or shell as we now call it, only has room for two electrons. So if we remove one of these electrons, electrons in the next shell will see the charge of the nucleus reduced by one unit, under the influence of the electron that remains.
6. In reality, the ions are not completely independent of each other. There is a statistical tendency for each ion to attract an excess of oppositely charged ions to its neighbourhood, and this lack of completely independent motion explains why the "van't Hoff factor", by which salt outperforms sugar, is a bit less than two.
7. Two words of caution here. On current thinking, there is some overlap between the Cl⁻ full shell (see below) and vacant orbitals on the sodium cation, and sodium chloride vapour consists of distinct NaCl units, although the electrons are very unequally shared and these units are usually referred to as "ion pairs", rather than as molecules.

8. For this, and much of the personal detail in this chapter, see Patrick Coffey, *Cathedrals of Science*, Oxford University Press (2008).
9. G.N. Lewis (1916), The Atom and the Molecule, *Journal of the American Chemical Society*, **38**, 762.
10. Scerri, p. 179; Richards, Nobel lecture, delivered 6 December, 1919, in *Nobel Lectures, Chemistry, 1901–1921*, Elsevier (1966).
11. Personal correspondence, quoted by Coffey.
12. Using modern terminology, the changes in free energy G, heat at constant pressure H, and entropy S for any chemical reaction are related by the equation $\Delta G = \Delta H - T\Delta S$. As T approaches 0, the difference between ΔG and ΔH obviously approaches 0. What is interesting is the fact that the difference in *slope* of plots of ΔG and ΔH against temperature also approaches 0. But this difference in slope is simply equal to ΔS. It follows that for any chemical reaction, the value of ΔS also approaches 0 as T approaches 0. This, essentially, is what the Heat Theorem says.
13. "Is" only with hindsight. In 1902, Lewis believed that the core of helium was an octet. In this belief he was joined by Mendeleev, who thought that there were light elements still to be discovered. In his 1916 paper, Lewis is uncertain whether helium is element number 2 or element number 4, although a proper appreciation of Bohr's work would have led him to reject the latter possibility.
14. Abegg spoke of "valence" and "maximum countervalence", and noted that these frequently added up to a total of 8. We can align Abegg's numbers, using current concepts, with the lowest and highest oxidation number of an element. Thus phosphorus shows a valence of 3 (and oxidation number of –3) in phosphine, PH_3, and a countervalence of 5 (and oxidation number of +5) in phosphoric acid, H_3PO_4. Similarly for nitrogen in ammonia and nitric acid (3 and 5), sulphur in hydrogen sulphide and sulphuric acid (2 and 6), and chlorine in hydrogen chloride and perchloric acid (1 and 7).
15. Why no diagram for oxygen, O_2? Because the Lewis dot structure in this case is misleading. For reasons that he could not have known, and that lie beyond the scope of a book such as this, the oxygen molecule actually contains three bonds, two half anti-bonds, and two unpaired electrons.
16. Some textbooks give the impression that hydrogen easily loses an electron, to form a cation, H^+, analogous to Li^+ or Na^+. This is not so. Such a cation would be a naked hydrogen nucleus, capable of

grabbing a share of the electron density in any chemical environment, and only capable of existing in high temperature or electrically excited plasmas. The so-called "hydrogen ion" that is so important in the chemistry of water, is actually more like $[OH_3]^+$, with the same electron arrangement as NH_3.

17. And, quite independently, the French chemist Joseph Achille Le Bel.

18. We can combine the orbitals in phase or out of phase. The in phase combination concentrates electron density between the atoms, where it is stabilised by both nuclei, and also increases the wavelength, thus lowering momentum (through the de Broglie relationship), and hence kinetic energy. The out of phase combination shows the opposite effects. It is sometimes necessary to consider how orbitals combine over larger groups of atoms, but two at a time is enough for our present purposes.

19. 1; 1 + 3; 1 + 3 + 5; later in the Table, to account for the rare earth elements, 1 + 3 + 5 + 7.

20. Suzanne Gieser, *The Innermost Kernel. Depth Psychology and Quantum Physics. Wolfgang Pauli's Dialogue with C.G. Jung*, Springer (2005).

21. Letter to A. Pais, 1950.

22. And just as the double seats on a bus will be occupied one at a time until passengers have to pair up, so will electrons distribute themselves in separate orbitals, until all the orbitals of a given energy have been populated. Moreover, just as passengers on a bus all prefer window seats, the electrons in this situation will line up their unpaired spins in the same direction.

23. Coffey, p. 307 ff.

24. Coffey suggests a large role for the enmity of Nernst and his associates.

Notes to Chapter 13 — Why Grass is Green, or Why Our Blood is Red

1. P.S. Braterman, R.C. Davies, R.J.P. Williams (1964), Properties of Metalloporphyrin and Similar Complexes, *Adv. Chem. Phys.*, **7**, 359.

2. Apart, that is, from H_2^+, formed in electric discharges through hydrogen gas.

3. This "linear combination of atomic orbitals" approach is pretty crude by today's standards. However, a few hundred lines of computer code will give you a pretty good solution to the Schrödinger equation

for two electrons moving under the influence of each other and of two hydrogen nuclei, and you end up with more or less the same kind of result.

4. B.M. Kayes *et al.* (2011), 27.6% Conversion Efficiency, a New Record for Single-Junction Solar Cells Under 1 Sun Illumination, *37th IEEE Photovoltaic Specialist Conference*, Washington, June 19–24, 2011.

5. Its compounds with carbon-containing species, especially carbon monoxide, are a partial exception.

6. The oxygen-generating reaction can be written as

$$4Mn^{3+} + 2H_2O \rightarrow 4Mn^{2+} + 4H^+ + O_2$$

where H^+ represents the hydrated hydrogen ion.

The high energy electron in the Photosystem II chlorophyll is passed down through a chain of molecules, some of them containing carbon-nitrogen ring systems very similar to that of chlorophyll itself, and in the process drives transfer of protons across the cell membrane, leading to the production of ATP (the universal short-term energy store for metabolism) from ADP and inorganic phosphate. When the (slightly different) chlorophyll of Photosystem I absorbs light, the high-energy electron is transferred to proteins containing iron-sulphur compounds, leading eventually to the conversion of $NADP^+$ (nicotinamide adenine dinucleotide phosphate) to its reduced (i.e. hydrogen-rich) form, NADPH. The latter is responsible for reduction of molecules that have incorporated carbon dioxide into sugar precursors.

7. Nick Lane's *Life Ascending* (Profile Books, 2009) gives an excellent and readable account of current thinking about exactly how this could have happened.

8. Strictly speaking, we refer to PSI and PSII in plants, and Reaction Centres RCI and RCII for the related structures in bacteria.

9. The other possibility is that PSI and PSII originated by the process of gene doubling, and then underwent a division of labour by diverging within a single organism. If so, bacteria now carrying only PSI or only PSII have lost one copy during their evolutionary history, retaining only the one more appropriate to their own needs.

10. S.J. Karpowicz, S.E. Prochnik, A.R. Grossman, S.S. Merchant (2011), The GreenCut2 Resource, a Phylogenomically Derived Inventory of Proteins Specific to the Plant Lineage, *J. Biological Chemistry*, **286**, 21427.

11. We parted company from chimps after chimps parted company from gorillas. So if chimps and gorillas are both "apes", and we want to use "apes" as the name of a clade, we must be members of it too.

12. For bees, see R. Menzel, M. Blakers (1976), Colour receptors in the bee eye — morphology and spectral sensitivity, *J. Comparative Physiology A,* **108**, 11. For vertebrates, see *The Making of the Fittest,* Sean B Carroll, Quercus (2008), especially Chapter 4, Making the New from the Old.

13. All, that is, apart from those life forms that derive their energy from geochemical sources, in places like hot springs or deep sea vents.

Notes to Chapter 14 — Why Water is Weird

1. This is not quite the simple tautology that it seems. We can arrange atoms in order of increasing electronegativity, and the degree of polarity in a bond will depend on the electronegativity difference between the atoms concerned. Thus the polarity of the bond in water is part of a more general pattern.

2. This is changing, and at an alarming rate; see Chapter 15.

3. The Crimean War, 1854–6, now best remembered for Florence Nightingale's work in establishing the profession of nursing, and the disastrous Charge of the Light Brigade.

4. By Le Châtelier's Principle, higher pressure will favour the arrangement that takes up less volume; see Chapter 6.

5. As you might expect, diamond also contracts on melting, but the situation is complicated by the fact that, except at very high pressures, the most stable form of solid carbon is graphite.

6. Using a combination of techniques to enhance the weak surface signal. See Stiopkin *et al.* (2011), Hydrogen bonding at the water surface revealed by isotopic dilution spectroscopy, *Nature,* **474**, 192.

Notes to Chapter 15 — The Sun, The Earth, The Greenhouse

1. I have simplified, but only slightly. For the outer 70% of the Sun's radius, heat is also transferred by convection. Near the surface, the hydride ion, H^-, takes over the role of major absorber and re-emitter of photons. However, hydride does have a high density of accessible

energy states, and functions in very much the same way as does plasma at greater depth.

2. See Chapter 5.
3. For a more rigorous discussion, see David Arnett, *Supernovae and Nucleosynthesis*, Princeton University Press (1996) Section 2.3. I have ignored the chromosphere because of the relatively small amount of matter that it contains, just as, later, I shall ignore the part of the Earth's atmosphere that lies above the tropopause, for the same reason.
4. R.A. Kerr (2009), Clouds Appear to Be Big, Bad Player in Global Warming, *Science,* **325**, 376, and references therein; A.C. Clement, R. Burgman, J.R. Norris (2009), Observational and Model Evidence for Positive Low-Level Cloud Feedback, *Science,* **325**, 460.
5. The effect is expressed most simply in terms of the equilibrium

$$CaCO_3(solid) + CO_2(gas) + H_2O \rightleftharpoons Ca^{2+}(aqueous) + 2HCO_3^-(aqueous)$$

6. Fourth Edition, Cambridge University Press (2009).
7. Bloomsbury Press, 2010.
8. http://blogs.telegraph.co.uk/news/jamesdelingpole/100092280/10-reasons-to-be-cheerful-about-the-coming-new-ice-age/, updated June 15, 2011; (accessed 24 Aug 2011).
9. G. Feulner, S. Rahmstorf (2010), On the effect of a new grand minimum of solar activity on the future climate on Earth, *Geophysical Research Letters,* **37**, L05707.
10. G.J. Retallack *et al.* (2006), Middle-Late Permian mass extinction on land, *Geological Society of America Bulletin,* **118**, 1398.
11. US Geological Survey, Volcanic Gases and Climate Change Overview, http://volcanoes.usgs.gov/hazards/gas/climate.php, (accessed 24 Aug 2011); Terry Gerlach (2011), *Eos* (house journal of the American Geophysical Union), **92** (24), 14 June 2011, p. 201.
12. National Science Foundation Investigation Closeout Memorandum, Case Number: A09120086, accessed at http://www.nsf.gov/oig/search/A09120086.pdf 24th Aug 2011; "Finding no research misconduct or other matter raised by the various regulations and laws discussed above, this case is closed."
13. With major perturbations, however, from ^{14}C released into the atmosphere in the 1950s by nuclear testing.
14. R.K. Kaufmann, H. Kauppi, M.L. Mann, J.H. Stock (2011), Reconciling anthropogenic climate change with observed temperature 1998–2008, *PNAS,* **108**, 11790.

15. A. Schweiger *et al.* (2011), Uncertainty in modeled Arctic sea ice volume, *Journal of Geophysical Research,* **116**, C00D06.
16. America's Climate Choices, report in brief, May 2011. The full report is available at http://www.nap.edu/catalog.php?record_id=12781 (free download, accessed 24 Aug 2011).
17. http://www.margaretthatcher.org/document/108237 (Accessed 24 August 2011).

Notes to Chapter 16 — In The Beginning

1. Lee Smolin, *The Life of the Cosmos*, Oxford University Press (1999).
2. Dirac was able to write the fundamental equations of quantum mechanics in a manner consistent with *special* relativity, but extending this result to *general* relativity, which includes an account of gravity, has so far proved impossible.
3. E. Hubble (1929), a relation between distance and radial velocity among extra-galactic nebulae, *PNAS,* **15**, 168; the preceding paper (M.L. Humason, The large radial velocity of N.G.C. 7619, *ibid.*, p. 167) credits Hubble with having suggested this test of the expanding Universe theory some time earlier.
4. G. Lemaître (1927), Un univers homogène de masse constante et de rayon croissant, rendant compte de la vitesse radiale des nébuleuses extra-galactiques (A homogeneous universe of constant mass and increasing radius, explaining the radial velocity of extra-galactic nebulae), *Ann. Soc. Sci. Bruxelles,* **47**, 49.
5. To be precise, in the *fine structure constant,* a ratio involving the charge on the electron, the electrical properties of free space, Planck's constant, and the speed of light.
6. Let H be the Hubble constant, which connects distance and relative velocity. Now consider two distant galaxies, at distances **r** and **R** from us, where **r** and **R** are what mathematicians call vectors (think of them as like arrows, with both magnitude and direction). So their velocities as measured by us are H**r** and H**R**, and their relative velocity is H(**r** – **R**). But **r** – **R** *is* the distance from one galaxy to the other one. So if there are astronomers on these galaxies, they, like us, will see themselves as being at the centre of an expanding universe.
7. See Note 4; also G. Lemaître (1931), The expanding Universe, *Monthly Notices of the Royal Astronomical Society,* **91**, 490; G. Lemaître (1931), The

Beginning of the World from the Point of View of Quantum Theory, *Nature,* **127,** 706.

8. B. Müller (2011), The Limits of Ordinary Matter, *Science,* **332,** 1513.

9. Including small amounts of ^{3}He.

10. E. Gawiser, J. Silk (2000), The cosmic microwave background radiation, *Physics Reports,* **333,** 245.

11. I have skipped over a complex series of events, since the proto-star needs to radiate away energy, otherwise the process will stop when the pressure of the heated, compressed gas balances gravitational attraction. The nuclear reaction sequences that I discuss are part of a much larger network, starting with the fusion of primeval deuterium to helium. For a review of these reactions, see G. Wallerstein *et al.* (1997), Synthesis of the elements in stars: forty years of progress, *Reviews of Modern Physics,* **69,** 995.

12. The only exception is ^{12}C, whose mass is set equal to exactly 12 units, by definition.

13. E.M. Burbidge, G.R. Burbidge, W.A. Fowler, F. Hoyle (1957), *Reviews of Modern Physics,* **29,** 547; often referred to simply as B2FH.

14. We can think of each of the extra neutrons as formed from a proton and an electron. However, the total number of low-mass particles (*leptons*) is conserved, so that if one low-mass particle (the electron) is consumed, another one must be created in its place.

15. This bald statement really applies only to the lowest energy level available, but all the other levels scale in proportion.

16. These occur in dense regions of active star formation, representing the final stage of stars that are massive and, as a result, short-lived. Type I supernovae occur in binary star systems, where a star initially close to the Chandrasekhar limit is pushed towards that critical value by matter that it has acquired from its partner. Strictly speaking, Type I and Type II are defined in terms of their spectra, and there are further sub-classifications, but that need not concern us here.

17. We now know that there are variations among Type 1a supernovae, and that comparisons should be restricted to those whose light emission shows similar decay curves.

18. T. Morris, P. Podsiadlowski (2007), The Triple-Ring Nebula Around SN 1987A: Fingerprint of a Binary Merger, *Science,* **315,** 1103.

19. To be more precise, a particular group of stars belonging to the Asymptomatic Giant Branch (AGB), during the third dredgeup stage: T. Lebzelter, J. Hron (2003), Technetium and the third dredge up in AGB stars. I. Field stars, *Astron. Astrophys.,* 411, 533; S. Uttenthaler

et al. (2007), Technetium and the third dredge up in AGB stars. II. Bulge stars, *Astron.Astrophys.,* **463**, 251.

20. The informed reader may notice a gap between lithium, formed in the Big Bang, and carbon, formed by the triple alpha process in stars. The missing elements, beryllium and boron, are formed when heavier nuclei are broken up by high-energy cosmic rays.

21. For example, an excess in some mineral grains of magnesium-26, formed by radioactive decay of the *s* process isotope aluminium-26, half-life 770,000 years, after that aluminium had already been incorporated in the grain. See also Note 15 to Chapter 1.

22. A.N. Halliday, B.J. Wood (2009), How Did Earth Accrete? *Science,* **325**, 44; J.F. Rudge, T. Kleine, B. Bourdon (2010), Broad bounds on Earth's accretion and core formation constrained by geochemical models, *Nature Geoscience,* **3**, 439.

Glossary

absolute zero: Temperature at which, in classical physics, thermal kinetic energy would be zero, although under quantum mechanics some non-removable *zero point energy* would still be present.

adsorption: Process in which a gas phase molecule is attached to a solid surface. First stage in heterogeneous catalysis.

albedo: The fraction of incident sunlight that a planet reflects back into space.

alloy: A solid material containing two or more different metals. Textbooks usually describe alloys as solid solutions, but they are better regarded as compounds, because the different kinds of metal element share electrons (see Chapter 12), and an alloy may even have a different crystal structure from the individual metals.

alpha decay: The process in which an unstable nucleus emits an alpha particle (helium nucleus), reducing the atomic number by 2 and the mass number by 4.

alpha process: The addition of an alpha particle to a nucleus, as in the conversion of ^{12}C to ^{16}O.

amine: Organic compound containing the $-NH_2$ or similar grouping bonded to a tetrahedral carbon atom.

amino acids: Molecules containing both the amine grouping and the carboxylic acid grouping; the building blocks of proteins.

ammonium nitrate: The salt, NH_4NO_3, obtained when nitric acid is neutralised with ammonia.

angular momentum: Momentum (mass x velocity) multiplied by distance (more precisely, perpendicular distance) from some centre. Angular momentum, like momentum, mass/energy, and electric charge, is perfectly conserved in all processes.

anion: Negatively charged *ion*.

annealing: Heating a material such as metal or glass to a temperature where the atoms can rearrange themselves so as to remove local strains and defects.

atom: The smallest amount that can exist of a particular element. The atom itself has a structure, with most of the matter concentrated in the *nucleus*, and most of the space taken up by the much lighter, but more spread out, *electrons*. The word "atom" means "unsplittable", but we now know that chemical changes rearrange the electrons, while *nuclear* processes involve changes in the nucleus.

atomic number: The ranking of an element in the periodic table, corresponding to the number of positive charges on the nucleus, which is equal, in a neutral atom, to the number of electrons, and controls chemical behaviour.

atomic weight: (More correctly speaking, "relative atomic mass") The mass of an atom, or average mass of an element, on a scale defined by $^{12}C = 12$; roughly equal to the average *mass number*.

Avogadro's hypothesis: The suggestion that (under the same conditions of pressure and temperature) different gases contain the same number of molecules. True to the extent that the *ideal gas* approximation is valid.

back reaction: The conversion of products back into reactants in a reversible reaction.

baryon number: Number of "heavy particles", i.e. protons plus neutrons. Conserved in all processes.

basalt: Non-acidic (i.e. low silica content) *igneous* rock.

base: Any material that can neutralise an acid.

beta decay: Process in which an unstable nucleus emits an electron and an antineutrino, leaving the mass number unchanged but increasing atomic number by 1.

biohydrometallurgy: Extraction of metals from their ore by microbial action.

bonding: The force holding atoms together to form molecules or solids.

Boyle's Law: Law stating that the volume of a gas is inversely proportional to pressure; when the pressure on the gas is doubled, its volume is halved.

Brownian motion: The random motion of small particles by chance as they are buffeted from time to time more strongly from one side than from another.

carboxylic acid: Organic compound containing the acidic grouping –C(O)OH.

catalyst: Something that is not itself consumed in a reaction, but changes its rate or course.

cathode-ray tube: An evacuated tube in which electrons from a heated negative electrode (cathode) produce a spot on a screen, as in a 20th century television set.

cation: Positively charged ion.

chaos: A situation where outcomes cannot be predicted, because every difference, however small, makes a difference.

Charles's Law: Law stating that the volume of a gas increases in direct proportion to increase in temperature.

chemistry of surfaces: What it says. The surfaces of solids are regions of high reactivity, because the surface atoms have unused bonding capacity, with important implications for catalysis.

chlorofluorocarbons or **CFCs:** Molecules containing Cl, F, and C only.

chloroplast: The internal organ of a plant cell that contains chlorophyll, and carries out photosynthesis.

classical physics: Physics as it was before relativity and quantum theory, involving absolute space and time, and precisely defined position and momentum for all objects.

combustion: Burning; rapid combination with oxygen, giving out energy.

compatible: (In geochemistry) Able to be incorporated into the same minerals.

composite: A material containing two or more chemically different kinds of solid component, intimately mixed.

compound: A substance in which two or more *elements* are held together by bonds. A compound is very different from a mixture. For example, chlorine is a gas, and sodium is a metal, but the compound sodium chloride is a *salt* — common table salt, in fact.

conduction band: See under *semiconductor*.

constructive margins: Plate boundaries where new *crust* is being formed.

convection: Process whereby heat is transferred by physical movement, e.g. by hotter fluid expanding and rising; watch a gently boiling saucepan.

covalent bonding: Bonding by sharing electrons, as opposed to *ionic bonding*.

critical temperature: Temperature at which the properties of a liquid and its saturated vapour become indistinguishable; above this temperature, no amount of pressure can lead to the formation of a separate liquid phase.

crust: Solid rock layer of the Earth, above the molten mantle.

decomposition: Reaction in which a compound is broken down into simpler components.

defect: An internal irregularity within a solid crystal.

deposition: Process of laying down rocks.

desorption: Process in which a molecule escapes from the surface of a solid into the gas phase. Last stage in *heterogeneous catalysis*.

destructive margins: Places where one plate is being pushed against another, with downward forcing of the *subducted* plate material.

diffraction patterns: Patterns of bright and dark that form when light passes through a regularly spaced arrangement of slits; any pattern formed from waves using the same principles.

dissociation: The breaking apart of a molecule into smaller fragments or, ultimately, atoms.

dynamic equilibrium: An situation where there is no overall change, although reactions are taking place, because the effects of forward and back reactions cancel each other out.

dynamite: Nitroglycerine stabilised by absorption on clay.

electrode: A pathway for electrons, specifically the place where electric current enters or leaves a solution.

electrolysis: Carrying out a chemical reaction using electricity. Overall, electrolysis always involves reducing some substance at the negative electrode, and oxidising the equivalent amount of another substance at the positive electrode.

electron: Negatively charged fundamental particle that is part of an atom.

electron degeneracy pressure: In highly compressed matter, pressure resulting from the fact that confining an electron more tightly in space increases its kinetic energy.

electron pair bond: A single *covalent* chemical bond, formed by the sharing of two electrons between two separate atoms.

electron spin: Not really spinning like a top, but representing the fact that an electron has a small magnetic moment, which can be "up" or "down" relative to a measuring magnetic field, but nowhere in between (quantum mechanics is like that; technically, the electron is said to have spin 1/2). Allows two electrons to fit into the same orbital, as long as they have opposite spins.

electronegative: Tending to pull electrons towards itself, so that in a covalent bond, the electrons will on average be closer to the more electronegative atom.

element: All of the millions of known and unknown kinds of substance are made up of various numbers of *atoms* of the 90 or so different *elements*, bonded together in different ways.

endosymbiosis: The process by which one organism makes its home in, and eventually becomes part of, another.

enzyme: Biological catalyst, usually protein-based.

equation: (In chemistry) The representation of a reaction using chemical *formulas* for the molecules involved. If there is more than one molecule of any kind, there is a number placed before the formula of that molecule to show how many of them there are.

equilibrium: (In chemistry) The point at which a reversible reaction is in balance, and will not show net movement in either direction. Equilibrium is often dynamic; both forward reaction and reverse reaction are in fact taking place, but at exactly the same rate, so that there is no net change.

equilibrium mixture: The mixture of products and reactants present at *equilibrium*.

erosion: Wearing away of rocks and soil.

exclusion principle: See *Pauli exclusion principle*.

fixed nitrogen: Nitrogen in the form of compounds that can be used by plants, rather than the free element.

fine structure constant: A ratio involving the charge on the electron, the electrical properties of free space, Planck's constant, and the speed of light.

flash photolysis: Generating short-lived reactive species rapidly using a brief pulse of light.

formula: A way of showing the contents of a molecule. In a formula, each symbol stands for an atom of an element. If there is more than one atom of an element there is a subscript after the symbol to show how many, and the symbols are joined together to represent molecules. A *structural formula* also shows how the different atoms are connected together.

forward reaction: The conversion of reactants to products in a reversible reaction.

fractional distillation: Method of separating gases or volatile liquids that have different boiling points.

free radical: Fragment of a molecule, with an unused capacity for chemical bonding to other fragments or atoms.

fugacity: The pressure of an ideal gas that would have the same chemical potential as a real gas.

galvanise: Coat with zinc.

Geiger counter: A device that counts the number of high energy particles passing through it, used to measure *radioactivity*.

geological column: Representation of the different periods into which we divide the Earth's history; oldest at the bottom, most recent at the top.

glycerol: (Common name glycerine) An organic substance, $CH_2OH.CHOH.CH_2OH$, obtained by treating fats or oils with alkali; hence *glyceryl*.

granite: "Acidic" igneous rock, containing silica as a major mineral component.

Green Revolution: Post-World War II revolution in agricultural production, doubling crop yields.

Haber (or Haber–Bosch) process: Process for the production of ammonia by the reaction between hydrogen and nitrogen.

heterogeneous catalyst: Catalyst of different phase from the reaction mixture, e.g. a solid catalyst for a gas phase reaction.

high explosive: Explosive, such as TNT or nitroglycerine, that contains oxidising and reducing agents in the same molecule.

hot body radiation: The radiation that any matter gives out by virtue of its temperature. Used here to mean *black body radiation*, the radiation emitted from a (hypothetical) body, which is a perfectly efficient absorber or emitter of light at all frequencies.

hydrogen bond: Interaction in which a hydrogen atom, carrying partial positive charge because it is bonded to a more *electronegative* atom, interacts with the lone pair of another electronegative atom.

ideal gas: A model gas in which the size of the molecules, and the energy of interaction between them, are regarded as negligible.

igneous rock: Rock such as granite or basalt that forms by solidification of molten rock forced up by volcanic activity into the Earth's crust (opposite: sedimentary rock).

indeterminacy: The idea that a clearly defined value for a physical quantity (e.g. position, momentum) may not actually exist.

inert: Chemically unreactive.

interference: Interaction between waves, reinforcing when they are in phase, and cancelling when they are out of phase.

intermediate: Something made in one step of an overall reaction or process, and used up in the next.

ion: An atom, or group of atoms bonded together, that has acquired electrical charge.

ionic bond: A bond formed by effectively complete transfer of electrons from one atom to another.

isoelectronic: Possessing the same number of electrons.

ionisation energy (or potential): Minimum amount of energy required to completely remove an electron from an atom or molecule. Often expressed in units of "electron volt", charge on electron x voltage, and this voltage is the ionisation potential.

island arc: Region where land is being formed through volcanic activity caused by the buoyancy of the lighter components of a plate that is being subducted.

intrusion: Rock forced into place between other rock that is already there; usually igneous rock forced up from beneath.

isomers: Molecules containing the same number of each kind of atom, but connected together in different ways.

isotopes: Atoms with the same number of protons, but different numbers of neutrons. Hence (almost) identical chemistry, but different atomic weights.

kinetic energy: Energy possessed by a moving object because of its motion, given by the formula $E = mv^2/2$.

kinetic theory of gases: Theory explaining the behaviour of gases in terms of the random motions of large numbers of individual molecules.

kinetics: The study of factors influencing the rate of a chemical reaction.

law of definite proportions: In any compound, the different elements are always present in the same ratio (for example, water always contains eight times as much oxygen by weight as hydrogen).

law of multiple proportions: If two elements form more than one compound, there is a simple relationship between the amounts of the elements in each of them (for example, while water contains by weight eight times as much oxygen as hydrogen, hydrogen peroxide contains sixteen times as much).

Le Châtelier's principle: The observation that if a change is applied to the conditions of an equilibrium, the equilibrium point will tend to shift so as to reduce the effect of the applied change.

lepton number: Total number of "light particles". Electrons and neutrinos have lepton number +1, while positrons and anti-neutrinos have lepton number −1.

line spectrum: The sharp lines of fixed frequency absorbed or emitted by an atom as it jumps between different energy levels.

lithophile: An element that is easily incorporated in the kinds of rocks found on the Earth's surface.

magma: Molten rock.

main sequence: A plot of temperature against brightness for stars, on which stars lie for most of their lives, while converting hydrogen to helium.

Manhattan Project: World War II US project to develop a nuclear weapon.

mantle: Molten layer of rock beneath the Earth's solid crust.

mass: Amount of matter. In practice, proportional to *weight*.

mass number: Total number of protons + neutrons.

mass spectrometer: Instrument that measures the mass of individual atoms (more strictly, ions) by using a combination of electrical and magnetic fields.

metallography: A way of examining the crystalline structure of metal, by etching a cut service so as to make the boundaries between different crystalline regions visible under the microscope.

migration: Movement of atoms or groups over a surface.

molecule: The smallest amount that can exist of a particular chemical substance. Most molecules contain more than one *atom* joined together, and dividing up a molecule would mean breaking the *bonds* between them.

momentum: Mass x velocity.

native: Occurring in Nature as the element itself, rather than as a compound.

negative (positive) feedback: A situation in which some change in a system tends to decrease (increase) the probability of more change in the same direction.

nerve gases: Poisonous gases that work by attacking the nervous system.

neutron: Uncharged fundamental particle that can form part of the nucleus of an atom, which accounts for the difference between nuclear mass and nuclear charge, and which decays outside the nucleus to give a proton, an electron, and an antineutrino.

nitric acid: HNO_3, the acid related to nitrates.

nitric oxide: Another name for nitrogen monoxide.

nitrogen monoxide: NO, also known as nitric oxide.

nitroglycerine: High explosive material formed from glycerol and nitric acid; technical name *glyceryl trinitrate*.

Nobel, Alfred: Inventor of dynamite; founder of the Nobel prizes.

nuclear: Involving the *nucleus*.

nucleus: (Of an atom) Small volume containing almost all the mass of an atom, in the form of protons and neutrons.

noble: (Usually of a metal) Unreactive.

opsin: A visual pigment.

orbital: A possible arrangement for an electron (considered as a wave) in an atom or molecule.

organic: A molecule containing carbon–carbon or carbon–hydrogen bonds.

organic peroxides: Organic molecules containing the highly reactive peroxide group; i.e. molecules of type R–O–O–H.

overall reaction (equation): Reaction (equation) that summarises a process that in fact takes place by way of a number of distinct steps.

oxidise: Combine with oxygen, hence, current meaning, remove electrons from an atom or group of atoms. Opposite, *reduce.*

ozone: O_3, an unstable form of the element oxygen.
  — **hole:** Area where the ozone layer is unusually thin or absent.
  — **layer:** Region in the stratosphere where ozone is formed naturally; the ozone formed there.

partial pressure: The pressure exerted by one of the gases in a mixture. The total pressure of the mixture is just the sum of the partial pressures.

Pauli exclusion principle: It is impossible for two electrons with the same spin to fit into the same orbital.

photodissociation: Process in which a molecule is broken apart by light.

photoelectric effect: Emission of electrons from a metal surface under the influence of light.

photolysis: Decomposition of a molecule by light.

photon: Light acting in the form of particles of definite energy; a *quantum* of light energy.

Planck's constant: the constant, actual value 6.626×10^{-34} Joule second, that relates the energy of a quantum of light to its frequency, and defines the scale of quantum effects in general.

plasma: A state of matter consisting of positively charged ions, and freely moving electrons. A plasma absorbs radiation of any wavelength, since the laws of quantum mechanics allow unbound electrons to have any amount of kinetic energy.

plate tectonics: Behaviour of the Earth's crust as a number of solid plates, formed at mid-ocean ridges or continental rifts and forced under each other elsewhere, on which the continents are carried.

poison (in catalysis): A substance so strongly adsorbed on the surface of a catalyst that it blocks the uptake of new material.

porphyrin: Any one of a family of molecules containing a ring of carbon and nitrogen atoms, with alternating single and double bonds, and the nitrogen atoms arranged in a square with their lone pairs pointing inwards. Examples include chlorophyll, and the haem proteins.

pre-Socratic: Philosophers before the time of Socrates and Plato. At that time, the activities of philosophy and science were not yet distinct, and many of the pre-Socratics seem surprisingly modern in their attempt to understand the mechanisms of Nature.

products: The substances produced in a chemical reaction, written on the right-hand side of the equation.

proton: Positively charged fundamental heavy particle (baryon) that can form part of the nucleus of an atom.

quantum (plural, **quanta**): The smallest possible amount of anything, especially the smallest possible amount of energy associated with light of a specific frequency.

quark: Component of a *baryon*. Protons and neutrons each consist of three quarks.

quench: (In metallurgy) Rapidly reduce the temperature, so as to trap the metal in its high temperature condition.

***r* process:** Process occurring in supernovae, whereby heavier atomic nuclei are built up by addition of neutrons, more rapidly than they can decay by beta-emission. See also *s process*.

radioactivity: The process by which some unstable kinds of atom break down (disintegrate), emitting particles and/or energy.

reactants: What goes into a chemical reaction, written on the left-hand side of the equation.

rearrangement: Process in which new bonds are formed and old ones broken.

red giant: Star at the stage of converting helium to carbon, or subsequent nuclear processes, after its core has run out of hydrogen.

red shift: Shift of a wave motion to lower observed frequencies, when the source and the observer are moving away from each other.

reduce: Convert an ore to metal; hence, post-Lavoisier, remove oxygen from a compound; hence, current meaning, add electrons to an atom or group of atoms. Opposite, *oxidise*.

relative atomic mass: Technically correct name for *atomic weight.*

respiration: The biologically controlled reaction of food with oxygen in the body.

reversible reaction: One that does not go to completion; a two way street reaction.

s process: Process occurring in red giants, whereby heavier atomic nuclei are built up by slow addition of neutrons, allowing beta-decay between additions. See also *r process.*

salt: A compound that consists of *ions.*

Scientific Revolution: The change in world-view between around 1550 and 1700, giving rise to our present idea of a Universe obeying uniform physical laws throughout.

semiconductor: A substance on the borderline between the metals and nonmetals, with a gap between a filled *valence band* of available electron orbitals, and an empty *conduction band*, that is not too great. Typically, its electrical conductivity increases with temperature, or with the controlled addition of impurities.

sequencing: Finding the order in which the information-carrying units of a biological molecule occur; for example, the order of A, C, G and T in DNA, or the order of amino acids in a protein. Similar sequences are evidence of common ancestry, and the amount of difference between two sequences increases with the amount of time since they last shared a common ancestor.

siderophile: An element than will readily dissolve in iron.

smelting: Reducing an ore to obtain molten metal (the words *smelt* and *melt* come from the same root)

solipsism: The view that the only thing that exists is one's own experience.

spectrum: A range of different frequencies of light.

steady state: Situation where material is being formed and destroyed at the same rate. This may either be an equilibrium, or a driven steady state that depends on the steady input of energy.

strata: Layer structures in sedimentary rock.

steps: The individual reactions that take place one after another to give an overall reaction.

stratosphere: Part of the upper atmosphere, between around 10 and 50 miles above the surface.

structural formula: Formula that shows how the atoms in a molecule are connected to each other.

subduction: Process by which crustal plate material is returned to the mantle at destructive margins.

theory: A suggested explanatory framework and its detailed mathematical or logical development. In this sense, the word "theory" does NOT imply that the explanation is still in serious doubt.

thermodynamics: The study of how energy is converted between different forms, such as heat, light, mechanical work, and chemical and electrical energy; the implications of this for chemical equilibria.

TNT: Trinitrotoluene, a *high explosive*.

to completion: 100% conversion of reactants to products.

tonne: Metric ton, 1,000 kg or 2,204.62 lb.

triple alpha process: Process in which three helium nuclei come together to form one carbon nucleus: $3\ ^4He = {}^{12}C$.

uncertainty principle: The principle according to which there is a limit to how accurately we can know information about any physical system. For example, the uncertainty in our knowledge of a particle's position, multiplied by the uncertainty of our knowledge of its momentum, must always be equal to or greater than $h/4\pi$.

unconformity: Place where sedimentary rocks lie over others that do not match, showing a time gap in between.

uniformitarianism: The idea that the laws and processes of nature were the same in the past, and will be the same in the future, as they are in the present.

valence band: See under *semiconductor*.

valence electrons: The electrons outside the core, whose behaviour determines chemical properties.

vector: Anything that has both magnitude and direction. Examples include difference in position, velocity, acceleration, and force, but not energy.

velocity: Rate of change of position. Velocity has two components, speed and direction. It is what mathematicians call a *vector* quantity.

wave mechanics: Mechanics using the wave-like properties of objects.

weathering: Physical and chemical processes wearing away the surface of rocks.

weight: The force exerted on an object by gravity. Thus your weight would be much smaller on the Moon, although your *mass* would be just the same.

yield: (Of a reaction) Fraction of reactants that get converted to products.

zero point energy: The amount of energy that remains in a bond and cannot be removed even at the lowest temperatures, because to do so would violate the uncertainty principle.

Index